LUCY'S KNEE

YVES COPPENS

LUCY'S KNEE
The Story of Man and the Story of his Story

Translation by Norman Strike

PROTEA BOOK HOUSE
PRETORIA
2002

Dr Francis Thackeray, Head of the Department of
Palaeontology of the Transvaal Museum in Pretoria,
acted as a consultant for the translation.

Lucy's Knee
Yves Coppens

Originally published as:
Le genou de Lucy
Éditions Odile Jacob, Paris, 1999
English translation by Norman Strike

First English edition, 2002
Protea Book House
PO Box 35110, Menlo Park, 0102
protea@intekom.co.za

Typography and design by Tienie du Plessis
Cover design by Tienie du Plessis
Reproduction by PrePress Images, Pretoria
Printed and bound by Interpak, Pietermaritzburg

ISBN 1-919825-89-4

© 1999, French text, Yves Coppens
© 2002, English translation, Protea Book House
© All rights reserved.
No part of this book may be reproduced in any form,
without prior permission in writing from the publisher.

*to Quentin,
the missing link, and to his Mum,
this first Quentintale*

This book is also dedicated – for everything they have taught me – to Christine's knee, Brigitte's elbow, Isabelle's hands, Serge's and Dominique's vertebrae, (another) Christine's pelvis, Valérie's trabeculae, Anne-Marie's legs, Yvette's feet, Renée's skull, Emmanuelle's and (another) Dominique's brains, Catherine's and Fernando's teeth, Pascal's mandible, Marc's collagen, Véronique's molecules ... and even José's discrete features.

It was written at la Cour des Prés and rue de la Digue, in Rumigny and Signy, but also in Nouméa and Paris.

Preface

It is no doubt normal to start a book, and especially this book, by offering the reader some sort of explanation of the title and subtitle we have chosen.

First the title. It obviously conjures up that charming part of the leg – where the bottom of the thigh meets the top of the lower leg – of the most beautiful Prehuman lady from the savannah of the Ethiopian Afar region, who had just turned 20 when, sadly, 3 million years ago, she perished by drowning in Lake Hadar. This Prehuman is Lucy, whose three co-fathers, Yves Coppens, Donald Johanson and Maurice Taieb, heading the expedition that discovered her, launched her on the international scene, making her the hitherto unrivalled star of the Great Origins Show. And the knee, indeed, is hers. Well preserved, this articulation of the femur with the tibia revealed – in a skeleton that told us that she was bipedal and walked erect – that her habitat was indisputably arboreal. There was no denying it: the lovely Lucy, who walked like a young model, still climbed trees like an old ape. Here we must salute the pioneering work of Brigitte Senut and Christine Tardieu, who came upon this revelation towards the end of the 1970s.

And so to the subtitle, which specifies that this essay is about the science that I try to serve (Chapters 1, 2 and 5) and the history of that science (Chapters 3, 4 and 6).

As for the two-tiered dedication, it speaks first of a Quentintale and then of spare parts.

A tale is a narrative poised somewhere between imagination and reality, and Quentin, as his name etymologically suggests, is "the fifth" hominid signed or co-signed by the author.

Mine was the only signature to *Tchadanthropus uxoris* (although it was recognized by Françoise Le Guennec-Coppens), probably a misdated *Homo erectus*, the first in the whole central part of Africa between its northern and eastern provinces; with Camille Arambourg, I co-signed *Paraustralopithecus aethiopicus*, the oldest robust australopithecine in East Africa; with Donald Johanson and Tim White, *Australopithecus afarensis*, the first australopithecine known to have possessed double locomotive capacity; with Michel Brunet, Alain Beauvilain, Émile Heintz, Aladji H.E. Moutaye and David Pilbeam, *Australopithecus bahrelghazali*, the first australopithecine on the "wrong" side of the Rift Valley; and, finally, with Martine Lebrun, I co-signed Quentin, the only one of the five to be born north of the River Loire. At the time of writing, the first four are between several hundred thousand and a few million years old, the fifth, between three and three and a half years old.

As for the spare parts, they are assembled in the bibliography.

Since I adore quotations and proverbs because they so marvellously illustrate the content they convey, in a kind of semantic shorthand, I have chosen one to open each of the six chapters of this book.

These are my sources.

A Chinese restaurant owner in the rue de la Montagne Sainte-Geneviève one day quoted me the first proverb. I immediately noted it down and asked him to write it out for me; but as neither he nor his wife, whom he consulted, were too sure of themselves, I had a colleague in Beijing write it out again; so it is from China that I have the present version.

The epigraph to Chapter 2 is a quotation from the *Odyssey*, which I heard again a few weeks ago in Ancient Greek and French from the mouth of Jacques Lacarrière, a Hellenist and writer.

The Senegalese proverb introducing Chapter 3 was chosen by Michel Egloff, founder of the Latenium in Neuchâtel, of which I am a trustee, to evoke the *raison d'être* of this new museum currently under construction. I am indebted to him for this superb expression of lucidity and wisdom which he so judiciously selected.

The quotation from the Bible was chosen by Dr Anne Sand for her Dor-le-Dor Foundation, "from generation to generation" – the audio-visual archives of contemporary Jewish history and memory – of which I am also a patron. I found it ideal to speak about my years of professional life and I have often borrowed it from my friend Anne, who gave me the text in Hebrew.

Finally, neither *Lucy in the Sky with Diamonds*, the refrain of the Beatles' song, nor the extract from Mr Pfister's letter require any comment.

I gratefully acknowledge all these sources of inspiration!

The book is divided into six chapters.

The first two represent my views – partly shared, partly not – on the Story of Man.

For the third chapter, which adopts the focus of some of my lectures under the same title since the 1970s, I am indebted to John Reader, a friend, for a number of historical references.

The fourth is my story; I actually used it to open the conference of the Paris Anthropological Society and the History

Centre of the University of Paris I "The History of Anthropology: People, Ideas, Moments", held at the Centre in the rue Mahler in 1989. The examples have consequently been drawn from the work of my own team or from neighbouring or associated teams.

The fifth is largely the result of work from my laboratory.

As to the sixth, it is almost entirely devoted to Lucy's influence in the French-speaking world. A worldwide collection of the poetic fall-out from the "Lucy effect" would undoubtedly swell the volume of material surveyed here; for example, I recently received a very pretty poem from a Czech writer, Miroslav Holub, to whom Lucy had also played muse.

My affectionate gratitude goes to Odile, Director of Éditions Odile Jacob, who ceaselessly encouraged me, had the boundless patience to wait a very long time for this text as well as the generosity to forgive my escapades – though never as an author – to other stables.

My thanks, also, to Anaïs Besnard-Statian, Monique Tersis, Marie-France Leroy and Christiane Doillon for carefully typing my successive and often barely legible manuscripts.

Thank you, again, to Christophe Boulanger, Michael Day, Valérie Galichon, Agnès Ménalte, David Pilbeam, Friedemann Schrenk, Herbert Thomas, Phillip Tobias, Erik Trinkaus, Michel Van Praet and Carl Voyer, to the second grade pupils of the Jules-Ferry primary school in Rheims, to Odile Jacob's team and to the museums of natural history in Geneva, New York and Paris, as well as to the Commonwealth Institute in London, for their assistance or permission.

And finally, of course, my thanks to Martine and Quentin whom I "love and adore", as Quentin says, and whom I robbed of so much time.

Contents

Chapter 1   "THE PREHUMANS"   *page 13*
The Story of Man before Man

Chapter 2   THE HUMANS   *page 33*
The Story of Man with Man

Chapter 3   CHRONICLE   *page 63*
The Story of the Story of Man

Chapter 4   AUTOBIOGRAPHY   *page 99*
The Story of My Story of Man

Chapter 5   LUCY THE FOSSIL   *page 123*
The Story of the Heroine of the Story of the Story of Man

Chapter 6   LUCY THE SYMBOL   *page 139*
The Story of the Story of the Heroine of the Story of the Story of Man

    Conclusion   *page 165*
    Glossary   *page 171*
    Bibliogaphy   *page 173*

CHAPTER 1

"THE PREHUMANS"
The Story of Man before Man
The Origin of the Hominids

树倒猢狲散

When there are no more trees, there are no more apes.

"The pre-humans"

Since Man* has been a conscious being, which he became between 3 500 000 and 2 500 000 years ago, he has been affected by the anxiety of wanting to know where he comes from, where he is going and what he is. All the origin myths of all human societies have been trying, ever since, to alleviate that anxiety by attempting to reply to its underlying questions.

And science, through its method of observation and interpretation, does no more than those myths have done. It tells us that Man, born of the living world, itself born of matter, on Earth, was born of the matter of the stars and of their long genesis through the expanding Universe. Man's situation thus appears to be one of immense humility. But science also tells us that that inert, omnipresent matter, became living matter on the Earth, then thinking matter, thus attaining by far the most advanced degree of complexity and organization known to us. And so Man's situation becomes one of immense importance. It is the brilliant manner in which science succeeds in unpacking this paradox that we shall endeavour to relate in the first two chapters.

Man is a living being. Life, the only one we know, is terrestrial and possibly Martian. Earth and Mars are both planets of a single system of a single star that we call the Sun. The Sun is part of a galaxy to which we have given the lovely name of the Milky Way. But this galaxy, itself already impressive in its dimensions and the number of stars it comprises – 200 billion, we are told by those who have counted them – is only one of 50 billion others which make up what we call the Universe.

* In this book the general term *Man*/*man* is used to include both males and females of the genus.

Man's story is therefore part of the story of Life, which is part of the story of the Earth, itself part of the story of the Universe – it is consequently a mere fragment of a single story. But it is possible, today, to tell 15 billion years of that story – History – to tell how that which is was. And why 15 billion* years? Because it is the age that we can attribute to the most ancient event in the history of the Universe that we are able to grasp for the moment.

At this point in time it is difficult to conceive of what happened before 15 billion years ago. However, we do know that at that time (an extrapolation, as the concept of a "year" is a construct based on the rotation of the Earth around its star, and the Earth would only be born 10 billion years later) – before time could be measured by the vibration of the caesium atom – inert matter existed, that it was composed of particles called quarks and was hot, very hot, dense, very dense, very elementary and terribly disorganized. But very soon this primordial matter (we must assume that it was the primordial matter until an even earlier, simpler one is discovered) would spread and at the same time grow more complex. Quarks would in effect organize themselves into nucleons, nucleons into atoms, atoms into molecules, and this whole little crowd would eventually cool and form galaxies, stars and planets.

Around 5 billion years ago – 4 billion 600 million years, we are told – the Sun and its system were born and the Earth settled at a remarkable (in the etymological sense) distance from its star – remarkable because, on the one hand, it allowed the abundance of condensing water vapour and the atmospheric gases which had accumulated to remain in their respective liquid and gaseous states, while on the other, it prevented them from breaking away from the Earth's

---

* The word *billion* is used to denote 1 000 000 000 ("one thousand million").

attraction, the planet's mass being sufficient to retain them. Venus and Mars were not to be as fortunate, the first being very large but too near the sun, therefore too hot, the second very small and too distant, thus too cold. And the Earth was hardly in full possession of its elements, themselves barely cool, when, in the depths of its basins, in the heart of the dark waters, inside material which facilitated their confinement, large carbon-based molecules started latching onto one another in chains, these chains clustering together in sacks and the sacks covering themselves in membranes – in other words, forming organisms. This step towards complexity and organization was immense.

The calendar, at this point, stood somewhere near the 4 billion year mark; some say 4 billion 200 million years. Somewhere in the Universe, matter, hitherto inert, had just sprung to life. Those units, constituted as we have just seen, and representing individuals, individualities, "persons" – nothing more nor less – were to start exchanging matter and energy between themselves and with the outside and assuming the power – the duty, we could say – to replicate themselves.

It would seem that in the environmental conditions prevailing at that time, matter had no option but to cross this threshold of complexity. One may therefore justly imagine that elsewhere in the Universe other planets, in similar situations to the Earth's in relation to their star, and possessing comparable environments to those which reigned on the Earth 4 billion years ago, were, are or will be able to experience the same transition.

This story, our own, thus reveals some of its main features from the start; it is factual, and closely linked to the environment, the fact being in effect environmental, in this case. Furthermore the story carries within itself a potential for increasing complexity that is ever ready to express itself, given the right time and the circumstances (the environment). But since nothing is stable in our constantly changing Universe, once the time and the circumstances (the environment) have passed, inert matter (the word *inert* is in fact very inapt to describe something which is hardly so) will never again give birth to living matter on Earth. The origin of life on Earth was a one-off event: before then, the conditions were not yet conducive to its emergence; afterwards, they were no longer so. Be that as it may, this was, for once, a matter of what was so loudly and rightly decried by our forefathers: spontaneous generation!

Life on Earth thus apparently had a single origin, in time and space; which means, little Quentin, that all living beings, without exception, which exist or existed, are related and are your cousins.

The story of life can be told as an immense family tree born of a primeval population of unicellular beings one day 4 billion years ago between a few layers of clay in the depths of one of the Earth's swamps. The grasses, the flies, the sparrows, the shrimps, but also the dinosaurs and the australopithecines are all part of our family; only the degrees of kinship vary.

In a very short time, the proliferation and diversification of life would be most impressive; it is true that it was to remain strictly unicellular for a long time, made up of cells with no nucleus for some 2 billion years, then of cells with a nucleus carrying a message for the transmission of a more personalized filiation during the next billion years. It was only at the beginning of the fourth and last billion years that, through the coming together of some of these cells, the first multicellular beings were formed; first vegetable, then animal – domains which were henceforth to remain differentiated. The first alliances also came into being in that period – the mitochondria that each of our cells protects are tiny beings which clung to our apron strings from then on; the chloroplasts of certain vegetable cells are others – and diversity became profusely inventive, thanks to the discovery of sexual reproduction in particular: "Qui fait un œuf fait du neuf," says André Langaney ("When an egg is laid something new is made"). Among other things, the first chordates were invented, equipped with the first spinal column, rigid yet supple, the framework of all fish and eventually of the mammals, including Man.

But with those unicellular beings, first without and then with a nucleus, those first multicellular beings of the static vegetable world and the (nearly always) mobile animal world, the first chordates, the first vertebrates, the first fish, we are still in the water – we have been in nothing but water for 3 billion 800 million years! Around 400 million years ago, first some plants, then some invertebrates, and finally a few vertebrates ventured out of the water into the air, for the very first time, on the terra firma of the continents entirely devoid of life until then.

The vertebrates of particular interest to us, from our chosen perspective, which is to end up with Man at the end of our journey, are the primitive amphibians, with their almost horizontal limbs. Some of them were to turn into reptiles, standing more raised on increasingly vertical limbs, and some of the reptiles into mammals, dividing into two the part of our story spent out of the water. Indeed, around 200 million years ago, the first of these mammals appeared, all of them oviparous, followed another 100 million years later by those which invented gestation in a placenta, along with viviparity. And the primates, all sorts of apes and monkeys, appeared in turn among the placental mammals to which they belong, around 70 million years ago, even though some authors, as usual, query whether the earliest among these creatures really are primates. Be that as it may, the earlier ones ate insects and the later ones insects and fruit, since flowering and fruit-bearing plants had also just developed. One man's meat is another man's poison, they say. And so it was that the preferred diet of our fathers, the primates, proved unpalatable to the dinosaurs, who enjoyed only gymnosperms (plants with exposed seeds) – now who was it who said that the dinosaurs were knocked out by a fatal meteorite?

And so the story goes on, wonderful in its coherence and inventiveness, each new development involving the position of the planet, the movement of the waters and the land masses, geography and climate, plant history, animal history, and their respective transformations (all of them undergoing these transformations only when forced to for reasons that we read in the cascade of events and facts immediately preceding such transformations). This natural history is indeed very much a story of events.

As a simple example, let us look at the period we have reached in our story and the order of the primates, which we have singled out: all the primates are tropical, and the very first of them appear in North America and Europe. (At the end of the Cretaceous period, a Euramerican continent separate from South America, Africa and Asia in fact lay latitudinally very far to the south of the current location of its present components, namely Europe and North America, now divided by the Atlantic Ocean.)

Their teeth with very low cusps, their brand new nails, their taut collarbones, their legs and forearms with separate bones, the extremities of their four limbs with five clearly individualized digits – everything about these new-born babes of the "creation" points to an arboreal habitat and a fruit diet. Indeed, at the end of the Cretaceous, the hitherto omnipresent gymnosperms had to give way to the conquering angiosperms (flowering plants with enclosed seeds): the new niche which thus resulted was not out on tender for long before it was filled by a jolly group who, from the outset, distinguished themselves by, among other features, enhanced three-dimensional colour vision, a tendency towards the quantitative and qualitative development of their central nervous system, as well as a social organization and a communication system vastly superior to those of their ancestors.

From one group of primates to another, from Euramerica to Asia and then to Africa and from Africa to South America, the exciting story of this group tells of their diversification and spread right up to the hominoids of the Eocene and Oligocene (40 million years ago), whom we believed to be exclusively Afro-Arabian until they recently popped up in Asia. Here once again, certain specialists, disagreeing on the grounds of the presence or absence of an external auditory meatus – how scientists do go on, with their fussy fixations – do not wish to see this limb of the tree branching out where it does, but higher up in time; there is no reason why they should not be right; the branching would thus be slightly modified, but not at all the thread of our story.

Later there would be many other hominoids, a veritable nursery of different forms, first in Africa – it seems – then in Eurasia; they have pretty names evoking places or resemblances and illustrating, as if it were at all necessary, the immense yet controlled fantasia of life as well as the rather more modest imagination of palaeontologists: *Proconsul, Dryopithecus, Ramapithecus, Sivapithecus, Gigantopithecus, Uranopithecus, Otavipithecus, Kenyapithecus, Morotopithecus* ..., the ape which looked like "Consul", a famous chimpanzee at the London Zoo in the 1930s, the oak ape, the Rama ape, the Shiva ape, the giant ape, the rain ape, the Otavi ape, the Kenyan ape, the Moroto ape. Every one of them, without exception, was at some time or another pronounced the ancestor of Man – each time with good reasons that, each time, would be contradicted by other equally good arguments.

The Proconsul-Kenyapithecus axis, however, has long enjoyed a degree of favour, with surprising persistence indeed, but its filiation and role as ancestor of the australopithecines, which some have sought so hard to establish, are far from being unanimously accepted. Let us say that these hominoids, both from East Africa, the first from the Lower Miocene (20 million years ago), the second from the Middle Miocene (15 million years ago), are where they should be when they should be; and as both are quadrupedal and tree-dwellers, but capable of occasional bipedalism on the ground (Anne-Marie Bacon), we can also say that they possess what they should have.

As for the autralopithecines – an ugly long name which has curiously become almost familiar to a wider public – they are, according to the knowledge currently at our disposal, the fossil hominoids closest to Man and come from the same biogeographical provinces as the first of the Humans – the first Men. As the oldest of them are, moreover, older than the oldest Humans, that suffices to make certain of the former the ascendants of the latter.

The same question can be asked about the first autralopithecines as was asked, we saw, about the first primates who lived some 70 million years ago or the first hominoids around

40 million years ago, namely what are the oldest fossils that can be considered eligible for inclusion in this category?

For the time being there are two that I would place in that privileged space: they were both found in Kenya, one in the Suguta Valley, in the hills of Samburu, the other in the Lake Baringo basin, near Lukeino; the first is between 7 and 8 million years old, perhaps a little older, the second, between 6 and 7 million; the first is represented by half of an upper jawbone complete with two premolars and three molars, the second by a lower molar (first or second). That obviously gives us very little to graft the prehuman branch of the family tree onto, but it is better than nothing, and all the more so since no fossil remains corresponding to the required profile have been found anywhere else in the world.

Moreover the presence of these possible hominids in East Africa fits well with the country which, as we have seen, has delivered many earlier hominoid remains and, as we have not yet seen, many later hominid remains. But it must be mentioned that some authors, through a somewhat different analysis of these tangible documents, attribute the Suguta jawbone and the Lukeino tooth to hominoids from before the birth of the prehuman branch, some sort of "pithecine" between the kenyapithecines and the autralopithecines situated before the crossroads. Martin Pickford and Hidemi Ishida have for example recently named the first of the two *Samburupithecus*.

As for me, I accord particular importance to those dates of 7 to 8 million years ago and, consequently, to the precious pieces of evidence coming from sediments of those ages, because they are not ordinary dates; they correspond to global astronomical and climatic events and to a whole cascade of local tectonic, climatic and ecological events. Globally, there was a crisis which has been well substantiated: the cooling of the planet and its consequent aridity in the tropical belts; locally, in East Africa, the rift phenomenon, which had been present for millions of years, was reactivated, resulting in subsidence and, all along the western lip of the great rift, the formation of mountains. The vegetation, on the whole very arboreal, which covered the continent from one ocean to the other, suffered an indisputable desiccation on the eastern side, which became more sparsely covered. It is probable that this hydrographic difference and the concomitant change in the vegetation between the east and the west of the rift, a difference which would intensify with time, went along with differences in the fauna and adaptations of this fauna, adaptations which had no reason to appear before.

Since, in contemporary nature, the living beings closest in every respect to Man are the great African apes (gorillas and chimpanzees), a proximity which can only mean shared ancestry, and since the great African apes, whose ancestors are unfortunately unknown to us, are to the west of this dividing line between the arboreal and the non-arboreal vegetation, while all the oldest (and I mean all the oldest) australopithecine remains are without exception to the east, an obvious scenario presented itself to me; or, rather, it forced itself on my mind one fine day in 1982, at a conference in Rome.

The common ancestors of Man and the chimpanzees must have lived there, in that equatorial Africa of savannah and forest; and then, circumstances having drawn a line from north to south down the middle of that equatorial Africa, a line which was to mark the separation between more humid and drier areas, these common ancestors found themselves divided into two populations with different adaptational constraints – alimentary constraints imposed by the circumstances and resulting constraints of stance and locomotion. In the arboreal zones to the west, the constraints of a diet based largely on fruits and seeds, even if some tubercles and game may be added; in the patchier less forested zones to the east, the constraints of a diet based on tubercles, roots and bulbs, even though one may add many fruits, more and more seeds and much more game (Fig. 3, map 1). In the first case, locomotion is arboreal, hence brachiate (swinging from the arms), and, on the ground, knuckle-walking, and the pelvis, among other parts of the skeleton, has the appropriate elongated shape; in the second case, locomotion is both arboreal and, on the ground, bipedal, and then exclusively bipedal, and the pelvis, among other parts of the skeleton, has the appropriate compacted shape; the pelvis, in addition to its locomotive functions and parturition, in effect now has to carry part of the body (Fig. 1).

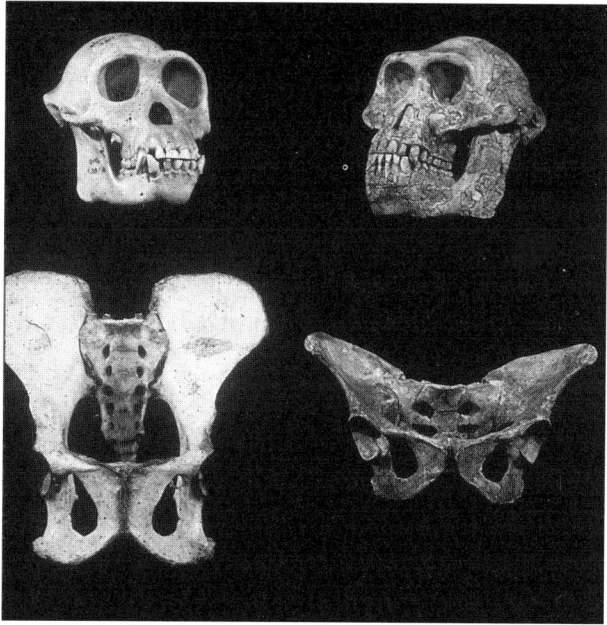

Fig. 1 – *Pan* and *Australopithecus*; on the left (to the west), skull and elongated pelvis of a chimpanzee; on the right (to the east), reconstitutions of the skull and the compacted pelvis of Lucy (document from the *Institute of Human Origins*).

This scenario or model, which I have called the East Side Story, is proving to be quite long-lived; it has the advantage of being simple, always a recommendable quality for a model, and of being supported by substantiated tectonic, climatic and ecological events.

Among its detractions, which are rare and "soft", a single one, to my mind, deserves attention (Louis de Bonis) because it is supported by experience that I in fact share. It could be expressed as follows: beginnings are discrete and difficult to locate because they are represented only by very small populations whose chances of fossilization are statistically weaker not to say nil. There is no more reason, in other words, to possess the base of the hominid branch than to possess that of any other branch whatsoever, such bases generally never being found, as it is.

Nevertheless, in terms of their possible role of representing the first hominids, the documents dating from between 6 and 8 million years ago which we have mentioned are reaffirmed by the discovery of other documents, once again in Kenya, still attributable to hominids, and which take over from them chronologically; these are half of a mandible between 5 to 6 million years old, found at Lothagam, southwest of Lake Turkana, and another half mandible 5 million years old, discovered at Tabarin in the Lake Baringo basin. From 4 to 5 million years ago, fossil remains become much more abundant. Those 4 400 000 years old, discovered at Aramis in the Middle Awash Valley in Ethiopia, have been attributed to a very particular form of australopithecine, so much so that it has been given a particular generic name in addition to its specific name: *Ardipithecus ramidus*. And then, from about 4 million years ago, two species of autralopithecines emerge at the same time, *Australopithecus afarensis* and *Australopithecus anamensis*, in sites in Ethiopia, Kenya and Tanzania; these forms were to lead a parallel existence for at least a million years in the whole biogeographical province of East Africa, before leading to totally separate destinies, between 2 and 3 million years ago – the first to a robust form of australopithecine, *Zinjanthropus* (*Zinjanthropus aethiopicus* and *Zinjanthropus boisei*) and the second to a new form of hominid, *Homo* (*Homo rudolfensis* and *Homo habilis*) (Fig. 2).

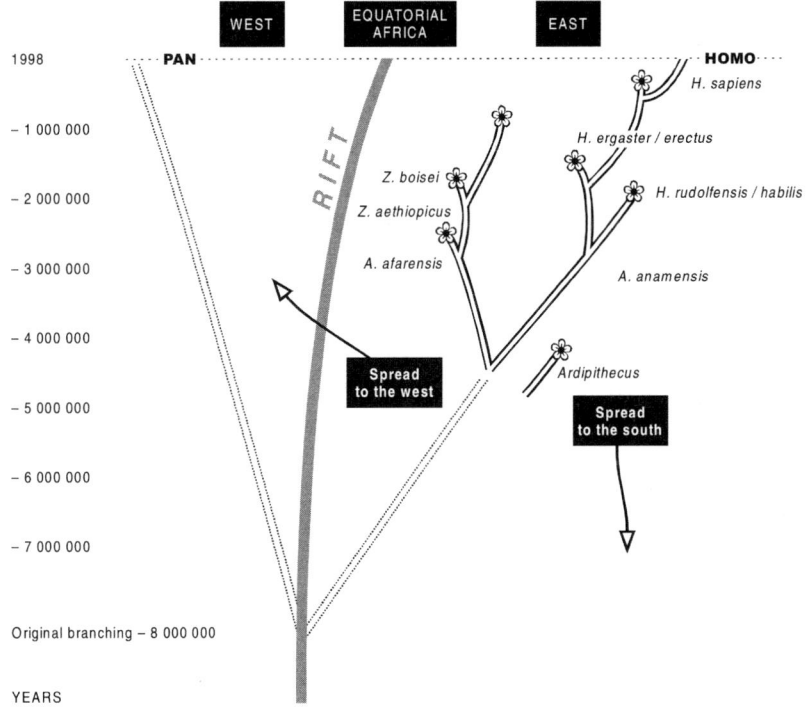

Fig. 2 – Phylogeny of the hominids

It was during the million-year coexistence of *Australopithecus afarensis* and *Australopithecus anamensis*, between 3 and 4 million years ago, that a probable expansion of the australopithecines' savannah habitat occurred, spreading from its East African nucleus in a southerly and westerly direction – rather more wooded towards the south, and grassy towards the west – and surrounding even better, like a halo, the forest nucleus curling round the Gulf of Guinea. This was the reason for the growth, in the south, of the *Australopithecus afarensis-Australopithecus africanus-Paranthropus robustus* cluster on the family tree, and for the sprouting of the *Australopithecus anamensis-Australopithecus bahrelghazali* side-branch, in the west.

Let us take the elements of this bunch of prehumans one by one in order to define their roles more accurately.

Nobody really knows what the documents assembled at Suguta, Lukeino, Lothagam, and Tabarin really represent; they are too discrete. The remains collected at Aramis, on the other hand, are sufficiently generous to enable one to start constructing the portrait of *Ardipithecus ramidus*, dating back 4 500 000 years. It is the portrait of a tree-dwelling (curved phalanges at the extremities of the limbs) biped (shortened base of the skull) with "primitive" (plesiomorphic) features (thin enamel, a single first milk molar) which are disconcerting if the remains are hominid, and even more disconcerting (given the location of their discovery) if they are not; if these creatures were hominids, it is strange that those ancestral features should have been maintained, at least 3 million years after their disappearance from other hominid lineages (Suguta, Lukeino …); if they were hominids and the features are newly derived (apomorphic), it could point to a new adaptation of hominids to an arboreal environment, a sort of regression (after all, the orang-utan provides a brilliant example of such a return); but if they were not hominids, we are perhaps dealing with large apes whose new adaptation to bipedalism (an apomorphic feature in that case) would be an unexpected and original innovation, with no filiation either with the earlier hominids or, a fortiori, with the later hominids, or with the great apes of the west such as we know them today.

The signatories of *Ardipithecus* (Tim White and his collaborators) do, however, see it as an ancestor of *Australopithecus afarensis*, no doubt because *Australopithecus afarensis* is the hominid that, as far as we know, comes just after it, chronologically. Moreover, Meave Leaky has just recently drawn attention to the astonishing resemblance between the arrangement of the milk teeth in *Ardipithecus*, which was both bipedal and arboreal, and *Australopithecus anamensis*, which was exclusively bipedal. It would no doubt be wiser, in the present state of our knowledge, to leave *Ardipithecus* where it is, with its great ape's silhouette or its possible status as the common ancestor of *Australopithecus afarensis* and *Australopithecus anamensis*, until we know it better.

As for *Australopithecus afarensis*, it is bipedal and stands upright (angle of the skull, curve of the spine, compacted form of the pelvis, obliquity of the femur), but it is also arboreal (instability of the knee and ankle joints, solidity of the shoulder, elbow and wrist joints, flat feet with the outer leading edge turned in, big toe turned away from the foot, curved phalanges of the extremities of the four members), rather like *Ardipithecus*. However, the resemblances with the latter stop there; for example *Australopithecus afarensis* manifests an incontestably hominid dentition (thick enamel, strongly molarized first milk molar).

Australopithecus afarensis appears thus as an ancient form from the still very wooded savannah, moving on the ground only for very short distances (the particularly wide pelvis implies a very energy-consuming waddling bipedalism). And yet, as an isolated divergent big toe was found at Sterkfontein, in the Transvaal (Phillip Tobias and Ronald Clarke), at a level probably a little higher than the 3-million-year mark (the oldest level for hominids in South Africa), one may imagine that this piece represents the first sign of the dispersion of australopithecines towards the south (recently confirmed by the discovery of a large part of a skeleton) and that it may be credited to *Australopithecus afarensis*, the possessor of this feature, thus making it the ancestor of *Australopithecus africanus*, then of *Paranthropus robustus*, born in that country which they never left.

And so Lucy might have been the explorer of Southern Africa, founder of a flourishing southern lineage, but also mother of a robust eastern branch; losing her role as mother of Man, she would thus have gained a double and original fertility.

Alongside *Australopithecus afarensis*, as we saw, another form of prehuman developed, from about 4 million years ago, called *Australopithecus anamensis* by Meave Leakey and her collaborators. On the evidence of fragments of a humerus and a tibia, *Australopithecus anamensis* appears to have been equipped with a poorly assembled elbow joint and a very solid knee joint, as are the humans of today, for example you, Quentin; *Australopithecus anamensis* could thus well have been the first inventor of this lineage with exclusive bipedal locomotion, without arboreal locomotion.

That is perhaps the *Australopithecus* that we encountered (AL 333) in Hadar (Brigitte Senut, Christine Tardieu, Dominique Gommery, José Braga), in the same sediments as the remains of *Australopithecus afarensis*. The hydrographic network, the source of those sediments, perhaps severely eroded the wooded savannah of *Australopithecus afarensis* and, to a considerably lesser extent, the neighbouring grassy savannah of *Australopithecus anamensis*; that would explain the far greater presence of the former and, just so that they would not be forgotten, the scant representation of the latter. It would also explain the obstinacy of many authors in considering all the prehuman material from Hadar as belonging only to a single species, that of Lucy.

Australopithecus anamensis, a possible ancestor of the genus *Homo*, could also be that of the West African australopithecine, whose broad premolars and molars, narrow symphysis (adjacent to the incisors) and deep palate bring it considerably nearer to Man (Michel Brunet and collaborators).

By drying out the eastern interior, the period of aridity that peaked around 4 million years ago thus entailed the dichotomous development of the prehumans into a bipedal and arboreal branch on the one hand, and an exclusively bipedal branch on the other (Fig. 2), and, by starting to dry out the exterior in a concentric circle from west to south around a forested nucleus, enabled this prehumanity to express, for the first time, its desire for expansion (Fig. 3, map 2) which would never cease to make itself felt throughout the following and successive humanities.

CHAPTER 2

THE HUMANS
THE STORY OF MAN WITH MAN
The Origin of Man

Ἄνδρα μοι ἔννεπε, Μοῦσα, πολύτροπον ...

Sing to me, Muse, about that Man of many tricks ...
Homer, *The Odyssey*

And then came another crisis, a major one, because it was the one that made Man.

For about 8 million years, as we have seen, the Earth, for reasons relating to its orbit around the Sun, has been steadily cooling down. Now these periods of cooling, which in the tropics result in spells of drying out, have occurred in irregular succession, in fits and starts, with pauses, surges, peaks and lows, and so forth, right until today. In the East African province which interests us, at the point that we have reached in our story, the climate has thus become steadily less humid and the vegetation ever more sparse throughout the last 10 million years; it is a tendency. I am insisting somewhat on this very clear notion of tendency, because, from time to time, some researcher, well-meaning but bent too closely over his analysis and taking the tiny part he is studying for the whole picture, declares that the forest had returned, that bipedalism was born under the trees, that the rainfall was more abundant than anyone had imagined, and comes up with various other circumstances too isolated to affect the evolution of the fauna. In this little book we have chosen to fly at an altitude from which it will be impossible to see these accidents of no consequence.

So we have this new peak in the cooling down process and in the aridity, that is to say the peak that I am proud to have brought to the fore in the 1970s thanks to the stratigraphic sequences of the Lower Omo River Valley, in south-west Ethiopia, and that I had therefore called the "(H)Omo event", the event which produced the genus *Homo* and which is revealed by the deposits emerging on the right bank of the river of almost the same name, an easy pun I could not avoid.

For many authors, this event dates back 2,5 million years, which is not wrong, but in fact, there too, a certain duration needs to be taken into account – from our perspective, namely for its effect on the fauna – beginning before 3 million years ago and peaking somewhere between 2,5 and 2 million years ago. The Omo deposits revealed to us that obvious correspondence between the appearance of the genus *Homo* (along with that of the "robust" australopithecine) and the evolution of the environment, simply because these deposits were the only ones exploited between 1960 and 1980 which generously illustrated that extraordinary period from 2 to 3 million years ago; the sites of the Olduvai basin were too young, the Hadar sites too old, and those to the east of Lake Turkana revealed a stratigraphical lacuna precisely at that age.

Since the Omo deposits, very rich in fossils and more than a 1000 metres from top to bottom, had the further advantage of offering an exceptional triple chronological check – biostratigraphic, magnetostratigraphic and radiometric – it was impossible for us not to draw some simple and obvious conclusions: from the bottom to the top of the sequence, that is to say from a little more than 3 millions years ago to a little less than 1 million years ago, the fauna and flora told that the weather had been changing, that it had become less and less humid. As "gracile" australopithecines, with tools, moreover, peopled the beginnings, and as they gave way to "robust" australopithecines – more and more robust – and to Men (the first humans) at the very moment when the ungulates were transforming their teeth to eat more grass than leaves and the ratio of the number of tree pollens to the number of grass pollens dropped from 0,4 to 0,01, it was difficult not to declare, at least in a whisper, that "it seemed very much as though the emergence of the genus *Homo* and of the robust australopithecine was in some way related to the drying out of their common cradle". Which is what I did in 1975. The assertion was duly decried, then borrowed ten years later, and actually became fashionable after 20 years, to the extent of becoming the favourite and, before long, a rather repetitive subject at almost every conference in our discipline at the end of the twentieth century.

Many methods that have nothing to do with palaeoanthropology but study the evolution of the climate have since confirmed the reality of this "(H)Omo event", albeit unwittingly – for example the measure of the ratio $^{16}O:^{18}O$ in the tests on foraminifera deposited on the floor of the Atlantic Ocean!

What exactly is this all about? In fact, it is no different from any other phylogenetic branching: before the crossroads, we have to do with species in equilibrium in their environment (*Australopithecus afarensis* and *Australopithecus anamensis*), then the environment changes; the species now in disequilibrium will consequently attempt to regain a new balance in the new environment imposed upon them and these "adaptations", if successful, will make new species of them or new genera (*Zinjanthropus aethiopicus* and *Homo rudolfensis*) according to the degree of transformation.

In East Africa, as we have seen, two species of australopithecines coexisted 3 to 4 million years ago: *Australopithecus afarensis*, the tree climber, and *Australopithecus anamensis*, the strider of the savannahs.

The great desiccation of the "(H)Omo event" then fell upon the land, probably throughout the entire tropics, in fact, and our family sought and found two responses to the crisis: the robust australopithecine, the *Zinjanthropus aethiopicus-Zinjanthropus boisei* line, with its daunting mass (1,50 m, 40 to 50 kg) and its nutcracker-like set of teeth (a rectilinear set of extremely reduced incisors and canines and almost parallel lateral rows of incredibly developed premolars and molars), and Man, the *Homo rudolfensis-Homo habilis* line or cluster, with his daunting reflective potential (brain size of 600 to 800 cc) and his omnivorous set of teeth (2 x 16 teeth of less differentiated dimensions and planted in a continuous curve).

This double response is absolutely exemplary, with, on the one hand, in the case of *Zinjanthropus*, a broadening of the vegetarian diet to include other harder types of plant to which the preceding australopithecines had no access and a larger body with no significant increase in brain size; and, on the other hand, in the case of *Homo*, an extension of the vegetarian diet to meat and an increase in brain size with no increase in body height or mass.

It seems to me – but I could be wrong – that, for the same reasons of adaptation, at least one of these two responses, namely that of the increased robustness of the australopithecines, emerged concurrently in Southern Africa but through no direct link with the East African robust forms. Thus, from a short *Australopithecus afarensis-Australopithecus africanus* line a robust line possibly developed within one and the same genus (*Paranthropus*), and perhaps within one and the same species, *Paranthropus robustus*, or through several, *Paranthropus crassidens, Paranthropus robustus*, and the even more robust and as yet unnamed *Paranthropus* from the new Gondolin site.

In East Africa, *Australopithecus afarensis* could be the ancestor of *Zinjanthropus*, and *Australopithecus anamensis* could be that of *Homo* (Fig.2); in South Africa, the same *Australopithecus afarensis*, in the process of transformation (*Australopithecus africanus*), could be the ancestor of *Paranthropus*. *Zinjanthropus* would thus be descended from a less evolved form than *Paranthropus, Australopithecus africanus* being so advanced that it has been taken more than once for the ancestor of the genus *Homo*.

On the *Homo* side, two questions of particular importance arise. On what authority may one speak all of a sudden of this new genus (or otherwise) of australopithecine? Does the genus *Homo* cover the concept, so philosophically loaded, of Man?

To reply to the first question, the palaeoanthropologist has to call on zoology. The zoologist says that, by definition, the minimal degree of difference between two animal forms which cannot interbreed to form fertile offspring is called the specific degree and the forms are called *species*; below that threshold we find subspecies or races, which can interbreed; above the threshold, the next most important degree of difference after the specific one is called the generic degree and the forms are called *genera* (plural of *genus*).

Since the palaeoanthropologist has no means of experimenting with, or even of merely establishing the interfertility of fossil forms, he must remain content to measure the degree of difference which exists between the forms he discovers and apply to that degree the corresponding zoological treatment together with its terminology. Between *Australopithecus anamensis* and *Homo habilis*, for example, the distance is sufficiently great to be labelled generic, in zoological terms; that is why the palaeoanthropologist calls the first form *Australopithecus* and the second *Homo*. Between *Homo habilis* and us, the degree of difference is incontestably smaller; as it corresponds to the degree which, in contemporary nature, is called specific, we bestow the same genus name on both hominids, but we attribute a different species name to each, *habilis* to the one and *sapiens* to the other.

Naturally there are authors who consider that the distinction should be situated elsewhere and who speak, for example, of *Homo africanus* instead of *Australopithecus africanus* and of *Australopithecus habilis* instead of *Homo habilis*. I suspect them of wishing to facilitate in this way the filiation demonstrations that hold their preference; in other words, I do not believe they are right. Indeed one could, using the arguments of geneticists, go to the extreme of climbing back up the lineages without ever changing the names of the representatives encountered along the way!

Between the privileged species of the genus *Australopithecus* that would give birth to the genus *Homo* and the first species of the latter genus, the distance for us is consequently that which separates two genera: the growth in volume of genus *Homo*'s brain, starting right with the earliest species of *Homo*, and everything that goes along with it (the complexity of the convolutions and of the vascular network of the dura mater, especially in the frontal regions of the brain) as well as the transformation of the teeth (their morphology, growth, dimensions, proportions and their organization designed for an omnivorous diet) essentially represent the distance we are concerned with here.

The reply to the second question, as to the synonymity of *Homo* with Man, is obviously negative; the definition of the genus *Homo* – a purely anatomical definition, as we have just seen – is far too restrictive. Some would like to characterize Man in terms of the consciousness that he acquires about beings, things and ideas, about himself, about others and about death – Man knows that he knows, they say – or in terms of the spiritual dimension of this same level of complexity, the emergence of the spirit (Gustave Martelet); others would prefer to distinguish him on the grounds of his ability to make tools, the tool testifying, as it were, to the existence of consciousness (Henry de Lumley); others again would stress a certain level of social organization – for these thinkers societies are, moreover, exclusively human (Michel Sakka). The last and the first groups would agree to add to their definition the ability to communicate through articulated language.

I believe that all these proposals are valid. But I also think that none of them falls within the province of the palaeoanthropologist. The emergence of consciousness, even if the stone artefact evinces it, can hardly be accurately dated; nothing tells us that the oldest tools discovered were the first and, in any case, nothing assures us as to who made what; the distribution of remains in the areas of hominid occupation only transcribes very poorly, for the moment, the social structure of which it is the imprint; and no objects have yet been found which have naturally recorded and preserved in memory the exchanges of ideas between australopithecines or early humans and, if it were possible – why not? – the means to read them. But since little flakes of quartz, jasper and chalcedony are found on sites dating back more than 3 million years in the Lower Omo Valley of Ethiopia and the tilt of the skull in the first humans suggests a simultaneous dropping of the larynx, one may almost say that the "(H)Omo event", 3,5 to 2,5 million years ago, was the circumstance which sparked off the flowering of novelties in this living world – consciousness, tools, society, communication – and made it a thinking world. It is surely not irrelevant to point out that all these elements, especially important in view of the development they were to undergo, have deep roots and do not all reach maturity at the same time. The definition of the genus *Homo* cannot therefore take into account any of the innovations listed above. As for the definition of Man, if it were to take them all into account, it would have to allow for the progressive emergence of philosophical Man, which started in prehuman times, before the emergence of the genus *Homo*, and continued for some time after the birth of the first species of the genus, in the time of the real humans.

We still do not know very well how "evolution" proceeds, but the relation between climatic changes and the transformation of species in the direction of their adaptation to the new environment has been established beyond doubt.

When, towards the end of the 1960s, I described the "(H)Omo event", I showed how it was accompanied by extinctions, departures, arrivals, inverted proportions and extraordinary local evolutions, all leading in the same direction; when one sees – the word is hardly an exaggeration – five lines of ungulates evolving all at once towards forms with higher molars crowned with more cusps, it is difficult to believe in the sacrosanct theory of chance mutations which will be retained by selection because, by chance, they happen to be better for the survival of the species.

And one finds oneself imagining, right in the very heart of the cell, in the individual arrangement of the chromosomes, a subtle mechanism which would be able to receive information from the changing milieu and use it, deliberately, to bring about the said changes, in the right direction. It would at any rate be more in line with the state of the situation in the field after the crisis, which bears no resemblance to a lottery result.

In that often considerable readjustment which follows a crisis, and which is obviously a consequence thereof, another fact emerges clearly: there is not enough room for everybody and, for those who are admitted to the "game", there will not be enough room for every single one of the respective proposals for adaptation they come up with. The new ecosystem that settles in has, like the preceding one, a limited number of places to fill and, consequently, can only admit a finite number of adaptations; the milieu having its own limits, the number of niches it can offer will not be infinitely multipliable; moreover, too many demands for a single niche, too many repetitive adaptations will give rise to competition and hence to losses.

This all means that, faced with a change of environment when it had been comfortably installed in its former environment, a species must react promptly and "know" that there will be no room for just any old "idea" and, indeed, no place for everybody. And so commences a race of ingenuity to invent a "trick" which, added to a species' inherited baggage – it will have to do – may suffice to pull through this time, provided that some other species, in the same turmoil, has not had the same idea just before or at the same time, in which case a battle will have to ensue, involving further risk. It must be said, however, that sometimes nature allows comparable so-called divergent solutions to exist side by side.

The point I am getting to is that the genus *Homo*, in order to negotiate the bend necessitated by the event of nearly the same name, perhaps did not "choose" the "large head and omnivorous jaw" solution, but that this option could have been imposed on it by the circumstances, and that the "larger body" or "toe-walking", or "herbivorousness" solutions, and so on, were pinched from it by others or were more difficult for it to effectuate genetically. Whatever the case, this "choice" was brilliant. The large head, in other words the greater brain mass, incontrovertibly endowed the genus *Homo* with superior reflective power; as for its omnivore jaw, it provided access to a broad opportunistic diet making it less vulnerable to shortages and enabling it to take far greater advantage than in the time of the australopithecines of the rich protein contribution from the carnivorous component of this diet.

And so here we have Man equipped, by definition, with greater curiosity (an obvious effect of reflection) and greater mobility (an evident result of his consumption of game and hunting, for even though Man did eat some carrion, he could not over-indulge in it, being physiologically incapable of consuming flesh that was too "high").

He is now endowed with better equipment; the artefacts which, it seems, preceded him, are ever more numerous and have gone on improving and taking on new forms in the hands of the genus *Homo*, which is inconceivable without tools (it is therefore not the tool per se which characterizes Man, but its permanence). Finally, because he has passed his adaptation exam, *Homo*'s demography is now definitely on the increase, even though this growth, given the small number of individuals at the start, was slow to gain momentum.

So, if one thinks of the first Men in these terms – higher primates whose numbers are steadily increasing, who are more and more inquisitive, more and more mobile and better and better equipped – it is not at all difficult to imagine them always on the move.

As to how this restless mobility worked, I see it as fairly similar to what I shall call the Inuit model: Robert Gessain and Paul-Émile Victor, who were in Greenland in the 1930s, told me that during the good years of abundant food and lower mortality, the little groups of Eskimos would grow until they reached a certain threshold number above which the normally accessible territory no longer sufficed to feed everybody; a small group – one or two families – would then break away from the original group, go and settle about 50 kilometres away, developing more or less rapidly until it reached the same threshold as before and then split and spread in its turn.

The spread of early Man could well have happened in like fashion, with accelerations in good years on good terrain and decelerations when conditions were less favourable, and probably even regressions occasionally. But after all, if one counts only 50 km of progress per generation, one obtains the modest figure of 15 000 years to get from East Africa to the farthest limits of Europe and Asia, and 15 000 years does not even represent the optimum resolution (margin of error) of any one of the absolute datings used today to estimate the ages of this slice of time from 2 to 3 million years ago.

I think that the genus *Homo* started expanding its territory right from its birth, for it manifested the behaviour associated with its characteristics from the very outset, precisely those which set it apart from *Australopithecus* and *Zinjanthropus*, making it well and truly the first species of a new genus. Since this birth can be dated to around 3 million years ago, I should not be surprised if *Homo* were to be discovered as early as 2 million to 2,5 million years ago at the western extremity of Europe as well as at the easternmost tip of Asia (Fig. 3, map 3). Tentative dates of about 2 million years ago in Europe and Asia are beginning to impose themselves with greater confidence; even earlier dates will no doubt be posited, until we bump into obvious buffer figures, as Man could not have arrived before leaving!

The nomenclature devised for early Man is somewhat confused today; one speaks of *Homo habilis*, *Homo rudolfensis*, *Homo ergaster*, *Homo erectus*, sometimes adding *Homo microcranous*, *Homo kenyaensis*, *Homo okotensis* (Valery Zeitoun). It is perfectly clear that there are too many.

Given that at the arrival point, anthropologists observe a great homogeneity in the present-day human species, to the extent that only a single human subspecies allegedly exists today – *Homo sapiens sapiens* – there is a chance that the original human species, the ancestor of the present one and which, in any event, was the first to spread, was unique.

If, on the other hand, a certain diversity can be confirmed at the dawn of the genus *Homo* and within the genus as a whole, it can perhaps be explained by the continuation, for some time, of conditions already observed in the prehumans, in other words, by a genetic differentiation under the influence of climatic and environmental variations and affected also by geographic isolation facilitated by the poor demographic density of the genus at the time of the first expansion attempts.

But if one snubs *Homo microcranous*, *Homo kenyaensis* and *Homo okotensis*, as they are not well substantiated, and if moreover one considers *Homo ergaster* as an old *Homo erectus*, one is left only with *Homo rudolfensis* and *Homo habilis* to commence the story, *Homo erectus* to go on with it, and *Homo sapiens* to end it – for the moment, naturally – (with the fascinating Neanderthal exception along the way).

And as some (Jean-Jacques Hublin) aver that *Homo erectus* is distinguished only by its archaic (plesiomorphic) features, and others assert that the transition from *Homo habilis* to *Homo erectus* and more so that of *Homo erectus* to *Homo sapiens* passes through all sorts of intermediary forms, one is forced to wonder whether it would not be better to see the story of the genus *Homo* as that of a single species – the human species. Its incontestable but continuous evolution would thus be exclusively constituted by successive degrees, a very particular biological evolution whose originality would obviously be due to the counteractive force that must have been exerted upon it by the cultural evolution which developed concurrently and, as we all know, with notorious extravagance.

Whatever the case, the genus *Homo* was now on the move; it increased its territory immediately, like an oil slick, reaching the Mediterranean and covering the whole southern part of Eurasia within a few hundred thousand years, and all this while still in one of its earlier forms (Fig.3, map 3). It is easy to understand that a *Homo erectus* born in East Africa 1,7 or 1,8 million years ago could not have "conquered" the world under that name and ended up in China, Indonesia, Georgia and Europe around 2 million years ago. It is time to correct certain books (even recent ones) and exhibitions by obstinate or out-of-touch authors; the shrewdest of them push *Homo ergaster* to the front of the stage to play for (geological) time, without losing *erectus* or face!

Since we pass, as we have said, from the first Man (*habilis? rudolfensis?*) to the second (*erectus*) and from the second to the third (*sapiens*) with no leaps, there is no reason for the second and third each to need a different cradle. I actually believe the first Man became the second wherever he happened to be at the time of that gentle slide towards less strong teeth and a more voluminous (but also thicker) skull, namely on the two continents of the Old World, Africa and Eurasia (the European peninsula is sometimes improperly treated as a single continent in its own right).

I think that, in the same way, the second Man became the third – with even weaker teeth, a skull considerably less thick and very much more voluminous – in the same place where the second arose, namely the two continents of the Old World or nearly: the whole of Africa and nearly the whole of Eurasia, less part of Europe and some Indonesian islands.

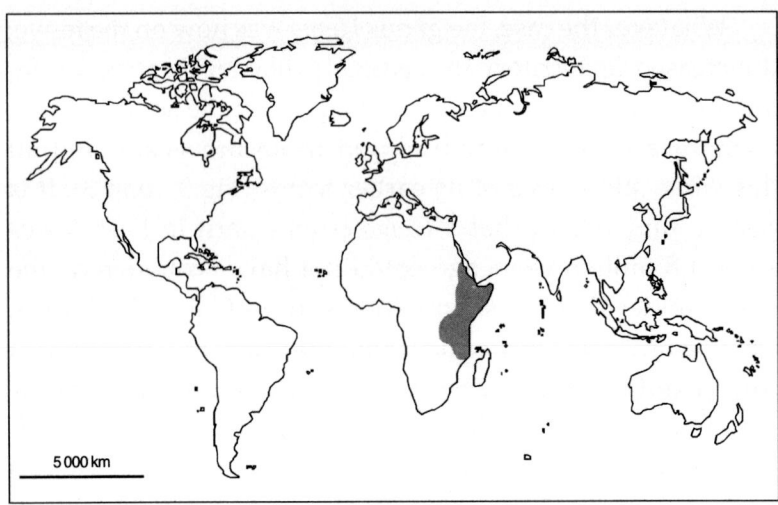

Map 1

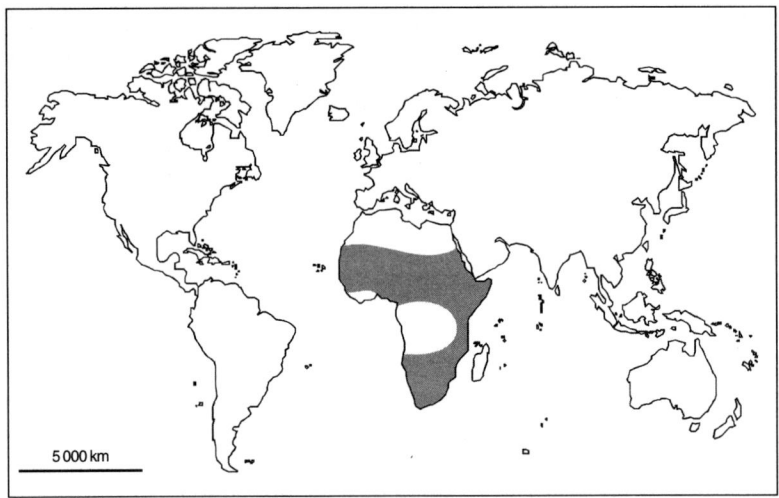

Map 2

Fig. 3 – History of Man's population of the Earth.
Map 1: 8 000 000 years ago; Map 2: 4 000 000 years ago.

THE HUMANS

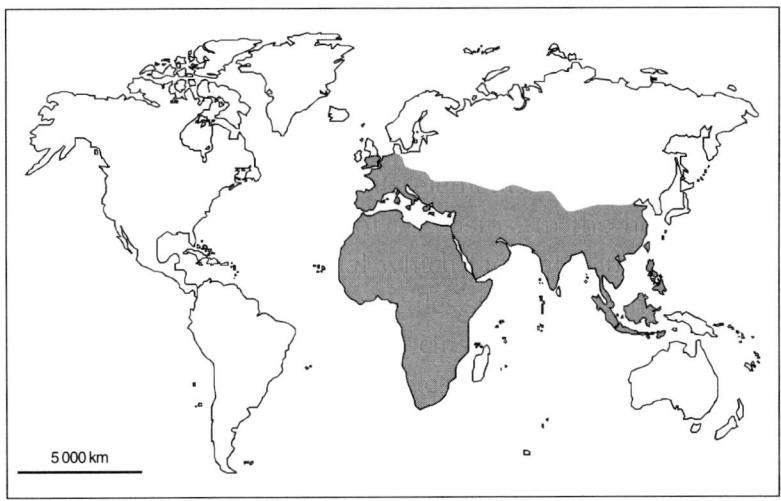

Map 3

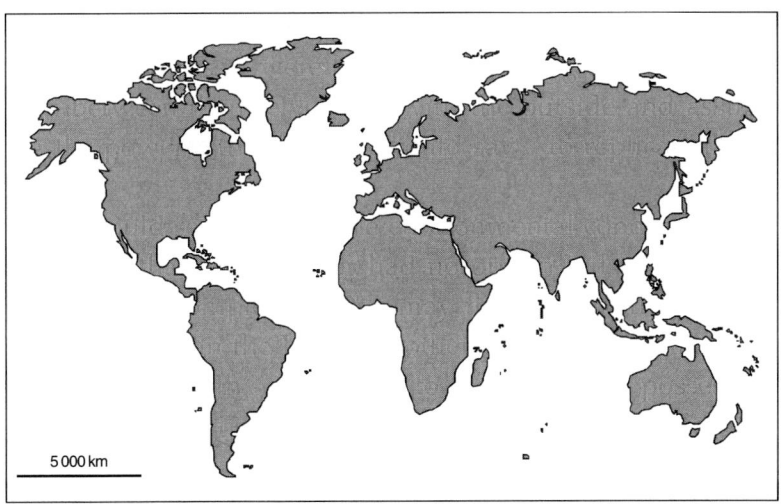

Map 4

Fig. 3 – History of Man's population of the Earth.
Map 3: 2 000 000 years ago; Map 4: 50 000 years ago.

We could in effect call Europe and Indonesia the exceptions to sapientization.

Let us imagine, 2 million years ago, the first waves of human expansion arriving in the western part of Eurasia, namely Europe, as well as in its eastern extremities, namely China and Indonesia. It is the time of the ice age, soon (according to convention) the dawn of the Quaternary period. Long cold periods separated by short warmer spells will succeed one another first irregularly then with an astonishing periodicity of 100 000 years. During the cold phases, the polar ice caps, the glaciers of the Arctic (including northern Europe), of the Alps, the Caucasus, Central Asia and North America will steal and lock up enough of the earth's water to cause a global drop in sea levels, uncovering the sea bed in places (the English Channel and the Java Sea, for example) and forming land bridges. The more temperate phases will obviously melt these extraordinary ice edifices sufficiently for the water to rise again and re-invade the uncovered areas.

In other words, Europe and Indonesia will alternately behave as islands; when Europe is closed off or almost closed off by the glaciers of the Alps and the Baltic and Scandinavian countries, Indonesia is attached to its Asian continent; when Indonesia is in turn cut off by the Sea of Java, the European peninsula is reunited to its Eurasian body.

The humans who had reached these continental extremities thus found themselves successively isolated from their mother populations in Asia and Africa, and then more or less reunited with them. But their isolation is more apparent than their possible contact during periods of bridging and continental attachment. These two populations in fact evinced biological behaviour that one could expect from isolated populations and which has been very well observed in vegetable and animal communities on all the islands of the world: they underwent a genetic drift or deviation – owing to the reduction of the space, the diet, the choice of genes – and this drift manifested itself in the appearance, on both sides, but more on the one (Europe) than on the other (Indonesia), either of autapomorphic features – characteristics which belong only to that form of humanity and are shared by no other – or of peculiar frequencies of features or associations of features.

So it is that on the geographically and genetically closed "island" of Europe, a superb drift developed, a model of its kind, that has been named Neanderthal, but which could just as easily have been called European. Neanderthal Man's autapomorphic characteristics are present in the most ancient human bones known in Europe (800 000 years old, at Atapuerca, in Spain) and they were never to leave the "European aboriginals" from 800 000 to 25 000 years ago, the date of their extinction – indeed, the same features are found on Neanderthal immigrants to the Near and Middle East around 200 000 years ago. These features would merely increase in number and frequency throughout those hundreds of thousands of years; first they were rare and statistically spread out – to such a degree that, in a given population, at the time of the first signs of the drift, certain individuals could be stamped with the Neanderthal seal and others not – then numerous and omnipresent in all the last populations to the extent that, for many authors, the only Neanderthals were those of the last 100 000 years. European Man from before that date would, according to those authors, be *Homo heidelbergensis* or *Homo antecessor*.

And that is how, in the Indonesian islands of today, at any rate on Java, which was incontestably closed off geographically and genetically, another drift developed (Jean-

Jacques Jaeger); not as lengthy and consequently less severe, this deviation may be called pithecanthropoid, after *Picanthropus*, the name given by Eugene Dubois to the first old fossil Man found in Java in 1891. It is interesting to follow the parallel and typically isolated evolution of certain great ungulates in Java, like that of an ancient elephant *Stegodon trigonocephalus* (Dirk van den Berg), right from its arrival on the island from the continent until the re-establishment of its contact with its congeners from the same continent, hence for more than a million years (from approximately 1 300 000 to about 30 000 years ago).

Neanderthal Man of the last 100 000 years was robust, but short in stature; he had a bulbous skull cap, with a receding forehead and chin, a massive brow ridge above high round eye sockets, a big nose in a puffy, muzzle-like face, its cheekbones lost in the mass, the teeth abnormally worn (Jean-Louis Helm).

To this "flattering" portrait should be added some other more technical but very characteristic signs: a small depression known as a suprainiac fossa, at the back of the skull, in the middle of the occipital bone, an orifice for the ear in the continuation of the cheekbone (which we call the zygomatic arch, and which looks like a handle on the side of the skull); and an empty space behind the wisdom tooth, a consequence of a certain projection of the face (prognathism) (Bernard Vandermeersch and collaborators).

The Java Man was taller than Neanderthal. He had a massive skull and the walls of the dome were particularly thick with prominent features (e.g. nuchal torus); the brow ridge was pronounced, the nose bones wide, the jaw robust, the teeth strong by comparison with what they would become in *Homo sapiens*. His million-year isolation enabled him above all to develop peculiar features on the parietal bones (sides of skull), the occipital bone (upper part), and the frontal bone (sinus) (Teuku Jacob). It is however during this long slice of time 3 000 000 to 50 000 years ago, that the genus *Homo*, the essential receptacle of thinking matter, acquires his dignity, in other words his paradoxical and sumptuous duality: freedom and responsibility. The characteristics of this thinking matter can of course be categorized in terms of consciousness, knowledge, cognition, ethics, aesthetics, creativity, intellectuality, spirituality, morality – all of them facets to which we have no access other than through *Homo*'s technology, his methods and the shapes and forms he produced. All we have to go by is our reading of a few remains, their distribution and the behaviours that can be deduced from them, the structures, the monuments, their decoration and the function that emerges. These means are indeed modest, but at the same time sufficiently eloquent for the story to be a superb account of the progress, both technical and symbolic, that they reveal.

André Leroi-Gourhan demonstrated ingeniously how tool efficiency progressed throughout a whole succession of given periods, by measuring the added lengths of the cutting edges – hence the active part – of flaked flints contained in one kilogram; in this way he arrived at the figure of 10 cm of cutting edge for 1 kg of flaked flint from 2 000 000 years ago, 40 cm for 1 kg of flaked flint from 500 000 years ago, 200 cm for 1 kg of flaked flint from 20 000 years ago and, almost incredibly, 7 000 cm for 1 kg of flaked flint microliths from 10 000 years ago. It is of course only an approximate quantification, but the exponential progression is sufficiently spectacular to have ensured the success of the demonstration.

Another author (Jean-Luc Piel-Desruisseaux) had some fun illustrating the extraordinary course of this exponential progression. He took a 24-hour clock (two clocks in fact) and considered each hour as representing 100 000 years, thus limiting the demonstration to the last 2 400 000 years. The first 12 hours illustrated the 10 cm of cutting edge per kilogram of flaked flint, the next eleven hours the 40 cm, and the last hour the 200, 2 000 and 7 000 cm of flint cutting edge but as well as the metal and steel cutting edges. One could say that a very long stalk thus preceded the incredible flowering that we are speaking about.

Having myself worked on the few human fossil remains discovered during Jean Chavaillon's excavations at Melka Kunture in Ethiopia, I thought I could add a certain demonstration of my own of the passage from living matter to thinking matter and to the progress that it brought about. The oldest levels represented *Homo habilis*, or at least his tools such as they were identified at the famous Olduvai site in Tanzania, where that species was described and defined for the first time; in other words, we were then in the slice of the 10 cm of cutting edge per kilogram of flaked flint. In the layers immediately above, *Homo erectus* began to stick his nose out – his humerus in fact; now this second Man happened still to be accompanied by the same tool technology as in the level below, from the same slice of 10 cm of cutting edge per kilogram of flaked flint. In other levels still a bit higher up, the second Man was still present, but now he knew how to turn out 40 cm of cutting edge per kilogram of flint tools. Immediately above we encountered the third Man, *Homo sapiens*, but associated with tools not exceeding the 40 cm of cutting edge per kilogram; the Man from the next layer up was still *Homo sapiens*, but with the record of 200 cm, and above that, again still *Homo sapiens*, with 2 000 cm, and so on.

In other words, we are seeing the unfurling of that long stem during which the genus *Homo* evolved first more quickly than his tools, and then the late and brief inflorescence of that stem characterized by a genus *Homo* which no longer changed but whose technology evolved in an astonishing burst.

This model, though full of defects, has the advantage of clearly showing that, in the story of thinking matter, it is first nature, in full swing, that is in the lead and stays there for a long time in the race with culture, before the relative speeds are inverted and culture forges ahead of nature. The model demonstrates at the same time how culture, in its continuous and inexorable development, came to respond more and more often in the place of nature to the demands of the milieu, simply because its response was quicker. The model in fact shows why and how Man's freedom was born: present in the first hominid who struck a stone in order to change its form to his or her advantage, this freedom has not ceased to grow, first in the shadow of nature, then openly, reducing nature's role, slowing down Man's biological evolution, erasing his instincts in favour of free will earned through conscious knowledge. It is knowledge – together with all that is conveyed by that word – that has made us free mammals; it is knowledge that will bring us an even greater freedom in the future, a freedom rising ever higher above the constraints imposed upon us by biology. It is a beautiful vision indeed that prehistory gives us of the future.

But let us return to that primordial gesture, so unremarkable and yet so fundamental, that stroke of one stone on another to change the form of the first. Whether the author of the act had intended and premeditated it, or simply done it with nothing particular in mind, or whether he or she had wished to reproduce the way in which a stone had accidentally broken, thus making the act more effective (Sacha Broussine), that inventor, to whom we, his or her grateful descendants, should raise statues, is the one who dared for the first time to cut nature's umbilical cord, act on his or her environment and change the world. And that hominid, by transforming an object, albeit a posteriori, naturally noticed the shapes – the shape preceding the stroke, and the shape resulting from it. That was 3 500 000 years ago or thereabouts, and the artisan was one of those prehumans that we call australopithecines but whose precise identity remains to be discovered.

The first tools described appear to some researchers to have been shaped somewhat haphazardly; to others they reveal already ancient experience. Chances are that all these authors are right, the first referring by comparison to later tools, the second implying the provisional non-existence of earlier ones. But little by little, the shapes of a certain number of tools would be standardized, for, after being tested over a long time for certain functions, they proved to be the best; and we have found them over hundreds of thousands of square kilometres where they were spread over hundreds of thousands of years, which means that they were transmitted through teaching, tradition and copying. As soon as toolmakers conceived definite ideas for the creation of certain products, they formed mental patterns or programmes to adapt these tools, programmes which were ever more elaborate and at the same time more and more numerous, for in addition to the increase in the number of centimetres of cutting edge per kilogram of flaked raw material, Man made great progress by increasing the variety of types of tools, the enhancement of his tool kit, a reflection of the growing number and diversity of his activities.

Around 1 800 000 years ago symmetrical artefacts start appearing to the west of Lake Turkana, in Kenya, as part of Man's set of tools. Needless to say that in addition to the progress achieved in the efficiency of the objects produced, this represents par excellence a considerable step in the observation of shapes and the ability to reproduce them. These objects, which are doubly symmetrical (both in face and profile), were to form part of the tool sets of hundreds of millions of people for far longer than a million years. One fine day 400 000 years ago, the famous so-called Levallois flake (after the Paris suburb where archaeologists first recognized it), namely the flake with the longest possible cutting edge, was struck off from the flint core, generously prepared by a programmed sequence of 13 or 14 strokes, in various regions of the Old World. The model of the paper hen produced by means of 13 folds, which keep one guessing until, with the fourteenth, a hen suddenly materializes, was used to great effect by Lionel Balout to illustrate the admirable abstract complexity of the Levallois blade.

Another aspect of the appreciation of shape and form, but probably also of colour and density, is provided by the surprising harvest in Europe of minerals and fossils of invertebrates in dwellings dating back a hundred thousand years. The collections of this kind really look like gratuitous assortments of curiosities gathered without the least intention of practical exploitation or any benefit other than intellectual or aesthetic. This long trajectory, 3 000 000 years of perception of ever more complex, premeditated and perfected – or appreciated and collected – shapes and forms, was in fact the perfect preparation for that explosion 40 000 to 50 000 years ago of sculpted, drawn or painted shapes and forms, created on walls or objects.

But progress is also the ability to project. It is probable that when an australopithecine sharpened a stone, it did so for instant use within minutes or hours. The object would have been shaped on the spot, with each utilization; it would not have been transported, or only rarely. It is probable that a *Homo erectus* who put the finishing touches to his biface or hand axe, after choosing the raw material and the colour sometimes as far as 40 kilometres from his home, and who, having used it and damaged it, took the trouble to reshape it and use it again, would once again have touched it up or reshaped it for further use within a few days or months. It is probable that a *Homo sapiens* who mixed his mineral or vegetable colours with clay and blood (bison blood has been detected in Upper Palaeolithic French cave paintings) to fix them, like one sprays a fixative on charcoal drawings to preserve them, went to all that trouble, which experience had taught him, to keep his paintings fresh for a few years or even a few generations. It is certain that the Man of yesterday who built the cathedrals or the Man of today who programmes flights to Mars or Jupiter knew, or knows, that he was or is working on the scale of a century or a millennium. And when the Man of science warns humanity against the extinction of the Sun after 10 billion years of existence while reminding us that it is already halfway through its life, he does no more than project himself into a certain future; and the successive dimensions of these projects grow exponentially, as we saw with culture in general and technology in particular.

Regarding the development of thought – first concrete and then more and more abstract and symbolic, with the evolution of language contributing massively – it is difficult to follow its trajectory other than from stage to stage, following the growing complexity of technology, and by attempting possible readings of rites, behaviours, signs and writing.

Between the mastery of symmetry and the achievement of the long preparation of the flint core, it is not impossible that the breaking of skulls (face and base), systematically observed in China and Indonesia, was intentional. It is certain that fire, encountered, observed and borrowed from nature for a long time already, was reproduced and maintained at about the same times, 500 000 years ago in the Far East (Zhoukoudian) and in the Far West of Europe (Menez Dregan). The burial of certain bodies, 100 000 years ago, in the East and the West, in each case accompanied by different coded signs – one must constantly bear in mind that in millions of years and millions of square kilometres the cultures are extremely numerous – represents a further and not inconsiderable step in the treatment of the existential anxiety inherent in conscious humankind. The message addressed to us by paintings and engravings from 40 000 to 50 000 years ago in Australia, Africa and Europe is the mark of an even more important increase in the growing complexity of the spirit. It is perhaps appropriate to point out here that, unlike all those who, sadly, see humanity's progress only in its knowledge and the applications thereof, prehistory brings an immense message of hope – and it is the only approach which, thanks to its broad time perspective, can do so – by demonstrating that since the emergence of consciousness and its product, reflection, Man's progress is quite as spiritual, ethical and moral as it is intellectual and technological.

And so, 50 000 years ago, *Homo sapiens* is everywhere in Africa and Eurasia, but not in Central and Western Europe nor in Indonesia, where Neanderthal Man and Java Man are happily evolving in the shelter of their glaciers or marine barriers.

Now a curious desire for expansion will once more take hold of this modern Man, and he will extend his frontiers in all directions and recommence his migrations, which had stopped around 2 000 000 years ago or nearly (Fig 3. Map 4).

Towards the north-east, the emersion of the Bering Strait with each glaciation will give access by means of a land bridge to uninhabited America, where he will spread very quickly from north to south (let us note that the same authors who refuse to see the genus *Homo* getting from Africa to Eurasia in a few tens, nay hundreds of thousands of years, are hardly surprised by the migration of this same *Homo*, admittedly better equipped, in a few tens of thousands of years or indeed a few thousand years, from Alaska to Tierra del Fuego, some 20 000 kilometres apart); towards the south-east, this Man will reach Australia, also uninhabited, after building rafts to get there; towards the south, he will once more set foot on Java, where he will find himself in the presence of an old occupant, Java Man, and towards the west, progressing through Europe, he will encounter Neanderthal Man, whom he already knew, in fact, for this European aborigine had gone in the opposite direction some time before and had settled in the Near and Middle East alongside him. In Java, we call this *Homo sapiens* Wadjak Man, because the first fossil modern Man discovered in Indonesia was found at the place of that name, in 1889–1890, and handed to Eugène Dubois for analysis; in Europe, the same *Homo sapiens* is called Cro-Magnon Man, because the first modern Man whose discovery attracted any attention is the one found in France, in a burial site, on the banks of the Vézère, in the so-called Cro-Magnon shelter, in 1868, during construction work for a railway line.

Curiously, 50 000 years ago also happens to be the time of that explosion, in several different centres, of what we today call art – perhaps a little earlier in Australia but we are not sure, and a little later in Africa and Europe.

The australopithecine knew a great deal, Man knew that he knew, and modern Man, in the full momentum of his conquering phase, had the desire to go even further than his predecessors and to make it known that he knew that he knew. And all of a sudden these hundreds of thousands of engraved or painted documents surge up out of time, before our eyes – surprising us time and again that our ancestors were not more stupid. Clearly these signs, whether in Australia, Africa or Europe, were at the time readable – I mean understandable – to at least some of those who had the privilege of seeing them; we are consequently speaking about a form of writing, even though it is not linear, which, much more cleverly than we think, associates the concrete signs we admire because we recognize them and the signs which we call abstract because we do not understand them.

And so it is modern Man who will brilliantly conclude this possession of the planet, prevailing over Neanderthal Man in Europe and Java Man in Indonesia, around 25 000 years ago in both cases, after several thousand years of coexistence with these curious, different and persistent other humanities.

The *sapiens* form of humanity subsequently experienced an astonishing development, as we know, first inventing, after 3 000 000 years of predatory economy, a new economy, a so-called economy of production, 12 000 years ago, then, with the industrial era, a particular form of production economy soon to be called consumerism, which has the merit of responding to the exponential demographic explosion that has marked the last two centuries, but, at the same time, the defect of depleting – consuming – the Earth's resources.

But above all let us not be afraid of the future. As we have seen, the peculiar characteristic of thinking matter – hence, possibly a prehuman and definitely a human characteristic – is this humorous paradox: we are *free* and incredibly strong because of this; we are *responsible* and terribly vulnerable if we forget that responsibility is the condition of our freedom. But since knowledge and, consequently, its transmission are the sources of the development of our freedom, we may to some extent declare that we do not have the choice of a policy, which is the last straw! We are free but not free to not learn!

And there we have modern mankind superbly engaged in the knowledge of its own body (the progress achieved in the mastery of fertilization or in that of the identification of genes is admirable); in the knowledge of its planet, which we can henceforth go around instantaneously in waves or electrons, or bodily in 90 minutes; and in the knowledge of its Universe. This increasingly generous understanding of space has made possible a certain grasp of time, of the time before and the time after, because, conveniently, this dimension has a direction. This inquiry has revealed the following information with crucial consequences for humanity: our star, the Sun, has only five billion years to live. We therefore have that much time to reflect, invent and organize in order to find the cultural parry to this natural aggression par excellence: either displace humanity or displace the Earth; but clearly both solutions will be feasible … in the long term.

Chapter 3

CHRONICLE
The Story of the Story of Man

*When you no longer know where you are going,
turn around and look where you have come from.*
Senegal

"As you grow older, you become interested in the history of the sciences you practise, or rather practised, because there is nothing else you know how to do. It is a well-known fact!" my friend and collaborator, Herbert Thomas, said to me one day about a friend. This was in the spring of 1991, in the sumptuous avenues of the park of the Marcelin-Berthelot station of the Collège de France in Meudon. I was thinking about my 1991–1992 course of lectures, which I had decided to devote precisely to the story of the story of Man, but he did not know this ... So, acquiescing with feigned warmth, I swallowed and he did not see; I said nothing but thought a while, or rather, to be frank, thought at length that he was not entirely wrong. At any rate, the remark had scored a direct hit; the precision of my memory of it is the proof.

So here then, in some sort of order, is that first contribution to the history of the sciences that I (still, for the time being!) practise; it is in fact more a collation of elements experienced, heard or read, to be added to the history of the palaeoanthropology file, than the work of a historian.

Nothing, of course, prepared us to meet fossil Man, and, even imagining that oral tradition or what is sometimes called collective memory had been able to cross several tens of thousands of years, why would it have specially retained and, faithfully or not, transmitted the "funny face" – according to present-day canons of beauty – that we must have had so long ago? In any case, transformations happen much too slowly on the scale of human life to have been able to mark people's minds and be inscribed in their memories.

It is therefore certain and natural that for the scientists of the nineteenth century, who were the first to take a rational interest in the history of Man even before relevant remains were discovered, our ancestors must have been beautiful; this notion of beauty, curiously synonymous with that of dignity, carried a heavy spiritual load. And there are still many people today who, although perfectly aware of the significance of the discoveries made, keep hoping that one day a "very ancient modern-looking ancestor" will be found; a very, very ancient fossil man looking very anatomically modern.

Although such a discovery could in the end do no more than push the same problem further back in time, without getting any nearer to solving it, one has the impression that to those people the circumvention of simian filiations would represent a real consolation.

The story of the discipline of palaeoanthropology was thus to unfold in such a climate, but, fortunately for its development, the discoveries were to be made in the inverse order of their age; first Neanderthal Man was discovered in 1829, then *Homo erectus* in 1891 and the australopithecine in 1924, with *Homo sapiens* or *habilis*, or indeed *rudolfensis* or *ergaster* only coming along to complete or to complicate matters. And despite this order, conducive to the advancement of ideas, we shall, in this third chapter, run through some incredible psychological obstacles to science.

It was in Europe that the urge to tangibly seek our ancestors first arose, but since it was precisely in Europe that the transformation of the genus *Homo* took a peculiar course, the first encounter with fossil Man, which proved to be Neanderthal Man, turned out to be something of a catastrophe! Neanderthal Man had in effect not only the receding forehead and chin of his (geological) "age", but he manifested also, among other "disagreeable" features, overdeveloped sinuses, whence a frighteningly large brow ridge above his frontal sinuses, and an "abnormal" bulging of his whole face above his maxillary sinuses.

It was Philippe-Charles Schmerling, a Dutch doctor of Austrian descent but practising in Liège, Belgium, who had the privilege of gathering the very first incontestable fossil Man remains. That was in 1829, in Belgium, in a cave named Engis, and the remains were those of a child's skull of what was thus later to be called Neanderthal Man.

Schmerling was surprisingly reasonable in his analysis and conclusions published in 1833–1834 in a work called *Research on the Fossil Bones Discovered in the Caves of the Province of Liège*, because he did not hesitate to speak of the contemporaneity between Man and the rhinoceroses, bears and mammoths found in the same cave: "I ended up concluding," he wrote, "that these human remains were buried in these caves in the same period, and, consequently, by the same causes which swept along a mass of bones of different extinct species." Moreover he intelligently completed this courageous account of the existence of Man before the Deluge with a demonstration of the presence of crafted tools in the same layers: "Even if we had not found human bones in conditions perfectly propitious for considering them as belonging to the antediluvian period, we would have been furnished with sufficient evidence by the carved bones and shaped flints."

It is a fact that the ideas of chronology associated with those of stratigraphy were in the air, but Philippe-Charles Schmerling's reasoning was nonetheless very modern for the period in its rigour and clarity. Unfortunately it cannot be said that this discovery was accorded the reception it deserved; indeed, important visitors like Charles Lyell, the Reverend William Buckland and Édouard Lartet came to inspect the pieces, the site and their discoverer, but they did not seem really convinced by what they examined. Only Édouard Lartet got away leaving the door open on the future, declaring in 1837: "There is nothing implausible about the idea!"

The second fossil Man discovery, again a skull, was made in Gibraltar in 1848 or shortly before, by quarrymen who were building the military fortifications of the famous rock; the first mention of the discovery actually appeared in the *Minutes of the Gibraltar Scientific Society* of 3 March 1848, under the heading "Human Skull from the Forbes Quarry, North Front". And there the matter rested. It was not until the visit to Gibraltar of an ethnologist, Thomas Hodgkin, in 1863, that the fossil was noticed and that an anatomist from London, George Busk, recognized it as being comparable to the Man of the Neander Valley, discovered in 1856, and of whom we shall speak now.

Once again it was quarrymen working in the limestone on the slopes of the deep Neander Valley through which the River Düssel flows, near its confluence with the Rhine at Düsseldorf, who discovered the remains of the "third" Neanderthal Man in 1856. These quarrymen had encountered a cave in the course of their work and had literally emptied it because it bothered them; they had then spotted some bones but, not recognizing them as human, had not paid them much attention and had reburied them with the debris of the cave. And it is to a teacher of natural science with an inquisitive disposition, Johann-Carl von Fuhlrott, that we owe the recovery, in August 1856, of some of the elements – the skull, a humerus, a femur, and a radius – of a skeleton which had probably been complete. Fuhlrott did what Hodgkin had done, he approached a professional anatomist to study them, and it was Hermann Schaafhausen, a professor in Bonn, who undertook the task; these remains were to become famous for they were to give their name to the first known, recognized and named form of human fossil, *Homo neanderthalensis*.

Like those of Schmerling, Schaafhausen's conclusions, expressed in the *Revue d'histoire naturelle* of April 1861, are intelligent, cautious and objective. "The bones of the limbs are exceptionally thick," he wrote, "and the muscle attachments are powerful; as to the form of the skull, equipped with a prominent brow ridge, it is natural and no doubt typical of this race." But these conclusions are of course embellished with commentary whose romantic style may surprise today: "It must be a member of those wild races from the north-west of Europe, barbarians whose appearance and fierce eyes must have terrified even the Roman armies," he wrote, for example.

In fact this was only the very beginning of the extravagant construction of a veritable myth of Neanderthal Man, a myth still very much alive today. Contributing to the construction were, on the one hand, the champions of the idea that the Man from the Neander Valley was an exceptional but isolated pathological case and, on the other hand, those who claimed that he was, on the contrary, a representative of a whole population which illustrated an ancient stage of human development.

Let us first quote some of the declarations of the famous defenders of the "Neanderthal individual".

August Franz Carl Mayer, a professor of anatomy at the University of Bonn, is probably one of the first, after his colleague Hermann Schaafhausen, to have ventured an interpretation of the remains of the Man from Neanderthal: "It is a degraded creature," he says, "which probably suffered from rickets in its childhood, which explains its bowed legs. But one also encounters that type of leg in those who spend their life in the saddle. It could therefore be a Cossack from

General Tchernitcheff's army, which camped in the vicinity before crossing the Rhine on 14 January 1814. These bones could be those of a deserter who had hidden and died there."

And Rudolf Virchow, an eminent pathologist, professor in Berlin and founder of the German Institute of Anthropology, confirms Mayer's arguments, taking them further: "These bones belonged to an elderly man, who suffered from rickets when he was a child, received serious wounds to his head in adulthood and had arthritis for a number of years before his death. With problems like those, he could not have survived among nomadic hunter-gatherers: he was thus a farmer belonging to a society where the elderly and sick were taken care of."

Another German anthropologist, who could not have liked his neighbours from the Netherlands, wrote: "His long low skull cap is typical of that of an old Hollander." And a French doctor, who, for his part, must have had a few scuffles with Breton neighbours: "It's a Celt who, like all Celts, has a low mental organization."

As for the evolutionists, although they very soon considered the Düsseldorf specimen as a type of extinct form of human, they did not conceal their discomfort when faced with the "brutal" appearance of its representative: *"agrioblemmatus"* it was called, for example, by Hugh Falconer in a couplet intended to describe the Gibraltar Man, which expressed, he said, both "the truculence of its eye and the savagery of its face"!

And unfortunately for the status and reputation of Neanderthal, this discomfort soon ceased; a dozen years after the disturbing discovery of the Man from the Neander Valley, the Man from the valley of the Vézère, Cro-Magnon, was discovered, and everything returned to normal, as many people had so fervently hoped, openly or secretly.

The Man from the collective burial site in the shelter under the rock of Cro-Magnon was, in effect, "beautiful"; he was tall (1,70 m to 1,80 m), had a very large skull, an impressive brain size (1 600 cc), an almost vertical forehead, a very modest brow ridge, a mandible with a pronounced chin, and so forth.

From that time on and for many, many years Neanderthal would be contrasted with Cro-Magnon, to an extreme degree, the former symbolizing archaism and crudeness, the latter elegance and refinement. Neanderthal had obviously become a lateral branch with no descendants, Cro-Magnon the only ancestor possible.

From 1886 onwards, discoveries of Neanderthal Men multiplied: that year there were those from La Naulette, then Spy in Belgium; in 1899, those from Krapina in Croatia; then, in 1908, those from La Chapelle-aux-Saints and Le Moustier; in 1909, from La Ferrassie; in 1911, from La Quina, in France, and so on. Those who had considered the Man from the Neander Valley as the representative of a widely spread human type and not as an abnormal individual were proven right by these discoveries: and this recognition was further concretized in a monographic report, still seen as exemplary today, on one of the skeletons of this type, the one from La Chapelle-aux-Saints, by the eminent French palaeoanthropologist and professor at the Muséum, Marcellin Boule. The report was published in Paris in the *Annales de paléontologie* in four volumes, the first in 1911, the two following ones in 1912 and the last in 1913.

But this increasingly documented recognition did nothing to alter the Neanderthal-Cro-Magnon contrast, which did not therefore mellow; on the contrary, it intensified to the extent of influencing the descriptions given by even the best anthropologists.

When Edmond Perrier, director of the Muséum National d'Histoire Naturelle in Paris, presented the La Chapelle-aux-Saints skeleton to the Académie des Sciences on 14 December 1908, he captured the atmosphere of what people thought of this Man at the time: "This fossil human type," he said, "differs from the present types and is situated beneath them, for in no present race does one find combined the features of inferiority that we observe in the skull from La Chapelle-aux-Saints. The Neanderthal, Spy, La Chapelle-aux-Saints group represents an inferior type, much closer to the anthropoid apes than any other human group. When, during the Upper Pleistocene, we find ourselves in the presence of individual manifestations of a higher order and of real works of art, the human skulls (the Cro-Magnon race) have acquired the principal features of true *Homo sapiens*, namely beautiful foreheads, large brains ..."

And Marcellin Boule, though he was a specialist, had not changed his tune five years later: "I must again point out," he wrote, "how much the physical features of Neanderthal, as I have just summed them up, are in harmony with what archaeology tells us about his physical aptitudes, his psyche and his manners. There is no industry more rudimentary and more miserable than that of our Man from Le Moustier. The use of a single raw material, stone, the uniformity, the simplicity and the coarseness of his tool kit, the probable absence of any trace of preoccupations of an aesthetic or moral nature fit well with the brutal appearance of the vigorous heavy body, the bony head with its robust jaws, which again reveal the predominance of the purely vegetative or bestial functions over the cerebral functions. What a contrast with the Men of the Cro-Magnon type, who had more elegant bodies, finer heads, straighter and larger foreheads, and who left behind, in the caves they inhabited, so much evidence of their manual skill, of the resources of their inventive minds, of their artistic and religious preoccupations, of their faculties of abstraction, and who were the first to deserve the glorious title of *Homo sapiens*."

And, in the same report: "[...] this conclusion [...] would show yet again that the latter (the Neanderthals) constitute a grouping or a branch which broke away very early from the human branch and that the evolution of the two groups started from a common core to which the lower apes are still relatively close."

Marcellin Boule was an excellent anatomist and his descriptions cannot be faulted, but, like anybody else, he received and was influenced by the ideas of his time and unconsciously adapted his interpretations to them: "Though not mechanically impossible," he wrote for example, "the total extension of the knee must not have been normal [...], the usual attitude must have been that of half-bent knees"; or this: "[...] altogether, what we know of *Homo neanderthalensis*'s spine reveals a certain number of features some of which may be considered primitive, such as the conformation of the cervical vertebrae, the strong development and direction of the spinous processes, the narrowness and shallow curve of the sacrum [...], it is impossible not to acknowledge that, with the morphology of this portion of his backbone, our fossil Man looks more like the chimpanzee than like contemporary Man taken as the term of comparison."

Now one of my masters, Camille Arambourg, himself a professor of palaeoanthropology at the Muséum National d'Histoire Naturelle – in fact the successor to the chair held by Marcellin Boule – told me the following anecdote. Wishing to visit a certain number of fossil sites in the Sahara, in 1948 he had organized a tour of the relevant sites thanks to the friendliness of some of the commanders of military outposts, which were French at that time, in the regions where the sites were located. He had thus hired a small aircraft in Algiers, had drawn up an itinerary and a time schedule with the pilot and had had his arrival announced by radio to these different outposts. On his first visit, the commander's car was there to meet him at the airfield; so Arambourg was able to get to the site, quite a distance from the airfield, with no difficulty; at the second outpost, things went equally well; but at the third, they landed before the car arrived. So Arambourg alighted from the little single-engine aeroplane and stood in the shadow of a wing to wait for the car. But the tyre on his side suddenly burst and the aircraft tilted to one side, not very severely, but enough for the unfortunate professor to be knocked on the head by the wing. The car arrived, the last visit was completed and the expedition concluded. Camille Arambourg returned to Paris, laden with an interesting harvest of fossils and notes, but went on suffering a little from an aching head and neck and decided to consult his doctor: the latter ordered an X-ray of the affected areas, and Arambourg then saw, to his astonishment, as one can imagine, that his cervical vertebrae were endowed with very large spinous processes oriented just like those of Neanderthal Man, features to which Marcellin Boule had imputed the permanently bent-kneed carriage of the Man in question!

Eleven years after Marcellin Boule's publication, Grafton Elliot Smith, professor of anatomy at the University of London, thus described "the uncouth and repellent Neanderthal Man" in a work titled *The Evolution of Man*:

> His short, thickset, and coarsely built body was carried in a half-stooping slouch upon short, powerful, and half-flexed legs of peculiarly ungraceful form. His thick neck sloped forward from the broad shoul-

Fig. 4 – "Improvement" of the image of Neanderthal Man (1909 left; 1990 right). (Left: reconstitution by Frantisek Kupka for *L'illustration* © Adagp, Paris 1999. Right: reconstitution by Michael Anderson under the supervision of Erik Trinkaus for the Maxwell Museum of the University of New Mexico.)

ders to support the massive flattened head, which protruded forward, so as to form an unbroken curve of neck and back, in place of the alternation of curves, which is one of the graces of the truly erect *Homo sapiens*. The heavy overhanging eyebrow-ridges and retreating forehead, the great coarse face with its large eye-sockets, broad nose, and receding chin, combined to complete the picture of unattractiveness, which it is more probable than not was still further emphasized by a shaggy covering of hair over most of the body. The arms were relatively short, and the exceptionally large hands lacked the delicacy and the nicely balanced co-operation of thumb and fingers, which is regarded as one of the most distinctive of human characteristics.

As for H.G. Wells, in his short story "The Grisly Folk", first published in 1921, he imagines Neanderthal, which has become synonymous with horror to the point of caricature, as follows:

> Hairy or grisly, with a big face like a mask, great brow ridges and no forehead, clutching an enormous flint, and running like a baboon with his head forward and not, like a man, with his head up, he must have been a fearsome creature for our forefathers to come upon …

Recent anatomy works, but also those on prehistory, and the new interpretations of the Neanderthal drift phenomenon have begun to restore the true image of this Man, but many prejudices still survive among professionals while they remain perfectly intact in the general public.

When I read the following passages, written in 1971 by Philip Lieberman and Edmond Crelin, whose competence I know and respect, I really cannot believe them:

> Neanderthal Man did not have the necessary anatomical equipment to produce the complete range of human language […]. He was not as well equipped for language as modern Man; his aptitude was however more advanced than that of the non-human primates of today and his brain must have been sufficiently well developed to have established a language based on sound signals.

I simply cannot imagine that a Man of 50 000 years ago (the one they refer to), who has so admirably mastered stone, who collected fossils and minerals, decorated his body with bracelets, necklaces and anklets, and who buried some of his dead whom he surrounded with multiple attentions, did not possess a language just as complex as our own. I really feel that we see here the perpetuation of a reputation whose resilience surpasses the rigour of interpretation – without otherwise affecting the quality of the analysis.

I am similarly disappointed by the caution of my friend, the Reverend Father Édouard Boné, when, in his comments published in 1978 on the Neanderthal burials – which offer, in deliberately dug trenches, bodies sprinkled with ochre, accompanied by functional or decorative objects, laden with food or covered with flowers – he declares:

> [...] it seems difficult not to read here [...] a certain attention to the corpse – however, this attention is perhaps only an expression of a certain form of affective closeness, no more [...], but, unless one gives to the words such a loose meaning and such a tenuous understanding that they become devoid of any real significance [...] we prefer to abstain from resorting to the metaphysical register when dealing with the Le Moustier burials.

In this case too, I cannot restrict myself to such a view; I simply cannot imagine Men, so close to us, for example, placing a block of limestone into which little cup-like hollows had been carved out in pairs, on the corpse of a young child, arranging three sharpened flints on the corpse of a foetus, digging three little trenches in the vicinity of these burials and putting in them animal bones, without some sort of symbolic ulterior motive and its ritual application.

As to my respected master, Jean Piveteau, he wrote in *Origine et destinée de l'Homme*, published in 1983:

> Neanderthal Man lived in a harsh climate [...], some mammals emigrated, others, like the reindeer and the mammoth, became numerous. And that represented new dangers for this Man.

and further on:

> [...] mostly, a south-facing position (of the dwelling) was preferred. Other necessities were no less decisive: the vicinity of waterholes and hunting grounds; the proximity of flint-rich terrains [...]. Although nothing permits us to conclude that Neanderthal protected his cave against incursions or attacks from the great carnivores, it is quite possible [...] that he closed the entrance [...]

and still further on:

> The manufacturing techniques of the Le Moustier industries were too complex to be transmitted through simple imitation: there must have been teaching, and by implication, language [...]. This Man was very skilful in skinning and butchering the animals he hunted. He must have had some fairly precise anatomical knowledge. There seems to be little doubt that the burial practices, whatever form they took, evince a sentiment that, while not absolutely identifiable with religion, remains related to it [..]. In the La Ferrassie deposit, six burials have been exposed [...], and this reunion in death has been seen as [...] the expression of a social life.

Clearly, with this text we have come very far from the fantasies of the pioneers, very far, in fact, from those of the anatomists of the first quarter of the twentieth century; however the romantic imagery has not yet disappeared completely. I cannot imagine that the Men referred to here were organized in any other way than in a structured society, rich in cultural and spiritual tradition, sufficiently in harmony with the environment not to have more problems with mammoths than we have today with wild boars, nor do I imagine them as anything but "smart" enough to have thought for a long time about preferably setting up their dwellings to face the sun, be protected from predators, near water, quarries and game, and to have acquired, just as long ago, sufficient ethnobotanical and ethnozoological knowledge to know how to identify the best berries and the tastiest game, in order to make excellent jam with the former and wonderful grills with the latter.

In other words, apart from the burials, these obvious and therefore futile observations would seem to me just as appropriate if they were applied to *Homo erectus* and *Homo habilis*, that is to say thousands, indeed millions of years before those for whom, as it happens, they are intended here.

As for Neanderthal's status among the broader public, it is still today so disastrous that it will take him many years to attain a more acceptable situation.

The naturalist Bernard Heuvelmans, a friend of mine, while travelling in the United States, one day at a fair "discovered" a character, real or a fake – thickset, its body covered in hair, its big toes turned out to face the opposite foot – placed in a large block of ice to preserve it ... or the better to conceal its nature and increase its mystery. He photographed it from every angle, described it as best he could but did not manage to acquire it, his curiosity having evidently bothered the "owner-exhibitor". Unfortunately the story was to end there, and Bernard Heuvelmans's proposal to provisionally name this being *Homo pongoides*, pending more detailed information on this exemplar or the acquisition of others, is perfectly defendable. What is not is to have titled the book which ensued *L'Homme de Néanderthal est-il toujours vivant?* (Is Neanderthal Man still alive?).

Furthermore, since all the hunters of yetis, almasties, sasquatches and big foots have for the moment only that fleeting reality to hang on to, all of them, or at any rate, all those I have met, are convinced that their quarry can only be Neanderthal Man!

The French film-maker Jean-Jacques Annaud, also a friend, decided one day to bring to the screen the famous novel by Joseph-Henri Buex, alias J.H. Rosny senior, *La Guerre du feu* (Quest for Fire) and, nearly 60 years after the book, he made a superb film in which, inevitably, superb prehistoric data (a little) and imaginary behaviour (a lot) are blended. But this is, to my mind, excusable in a work that does not claim to be a document. What can on no account be considered such is the filmmaker's commentary when he situates the action of his film 80 000 years ago (one wonders why) and calls his characters Neanderthals and Cro-Magnons.

I wonder to what extent the long-asked question – and which has perhaps since been answered in the negative – about the possibility of a cross between Neanderthal and Cro-Magnon did not partially derive from similar presuppositions. There was significant opposition when, in the 1930s, the discovery at Skhul, in what was then Palestine, of skeletons whose skulls had large foreheads, pronounced chins but persistent brow ridges, had made the discoverers think that they had found hybrids between Cro-Magnon and Neanderthal: if such hybridization had really taken place, some said, leaving the possibility open to future research, it must have been very rare and could not have affected the main stream of human evolution; such marriages, others dared to declare, must only have given birth to sterile descendants, like mules!

Despite the persistent memories of their former situation, the Neanderthals have however improved considerably as they have grown older (Fig. 4)! Certain specialists are starting to take them for what they are, really civil beings, educated, sensitive, curious, artistic, technicians and thinkers, musicians and philosophers, and who, clothed in well-cut dresses or suits and wearing scarves or well pulled-down hats, would not be noticed in a crowd today, not even by anthropologists.

The second stage was also to start in Europe at the end of the nineteenth century, but would be played out in the Far East, first in Indonesia, then in China.

A young Dutch doctor with a good French name, Eugène Dubois, was convinced at that time that the ancestors of the ancestors then under discussion, namely the Neanderthals and Cro-Magnons, must have lived in the tropics for, he thought, the Europe of the glaciers could not have been able to shelter them, for climatic reasons. This was in Amsterdam in the 1880s, and Dubois, a fervent reader of the German Ernst Haeckel, who already used the name *Pithecanthropus alalus* (the ape-man with no language) to refer to this theoretical ancestor living on a vanished continent which had occupied the place of the Indian Ocean, Lemuria, champed at the bit as he wondered how to get to those far-off regions.

Astonishingly determined, he found a solution by signing up as a doctor in the colonial army, in this case the army of the Dutch East India Company, which embarked him with wife and child (he had a little girl) for Sumatra in the autumn of 1887.

He remained on that island for a number of years, prospected a great deal, discovered numerous fossil vertebrates whose species he shrewdly dedicated to his superiors, then contracted malaria.

It was while convalescing from this illness that he was sent to Java in March 1890, a new terrain where he organized further prospecting, opened new sites, discovered new fossils and continued to dedicate some of them to his officers so that they would leave him be, which is what they did. He started excavating in caves as he had been accustomed to doing in Europe, then switched to exploring open-air sedimentary sites. In November 1890, he picked up the first fragment of an ancient mandible at Kedung Brubus, which he called "human but of a different and probably inferior type to existing humans and to the extinct antediluvian species". And in September and October 1891 he successively harvested a tooth and then a skull cap at Trinil, on the banks of the River Solo, fossils which he announced and interpreted as follows: "The Pleistocene fauna of Java, which was increased by a chimpanzee molar in September this year, was again enriched a month later. Near the site where the molar was found, on the left bank, a beautiful skull cap was excavated; it belonged, without a doubt, like the molar, to the genus *Anthropopithecus troglodytes.*"

Nearly a year later, in August 1892, a femur was extracted by the diggers from the same lapilli formation and, in October, another isolated tooth. Impressed by the human character of the femur, Eugène Dubois first changed the species name of his fossil while retaining that of the genus; thus in 1892 he called it *Anthropopithecus erectus*, the upright chimpanzee. The following year, contemplating this really very human femur and this very large skull cap, he created or brought to life Haeckel's *Pithecanthropus*, but calling it *Pithecanthropus erectus*. "It was," he wrote in an unpublished manuscript, "a matter of the animal most resembling Man and clearly forming the link between the latter and his closest mammalian parents, as the theory of development supposes, the transitional form which, according to the teachings of evolution, must have existed between Man and the anthropoids."

At the end of his eight-year contract in the army, Dubois returned to Holland; this was now in 1895, and it was at the Third International Congress of Zoology, meeting in Leiden that year, that he presented his fossil remains (the originals) – and his interpretation of them – to the scientific community for the first time.

And during this second stage we shall once more encounter that same reticence, which was to grow more or less consciously into resistance and rejection, that we came up against in the course of the heroic story of Neanderthal Man.

The participants at the Leiden congress, moreover under the chairmanship of Rudolf Virchow, saluted the incontestable interest generated by the discovery, but were not in the least convinced of the status attributed to it by Dubois, and the same scepticism would surface at subsequent scientific meetings in Liège, Brussels, Paris, London, Dublin, Edinburgh, Berlin and Genoa.

Eugène Dubois was extremely annoyed by these reactions; he ascribed the criticism of his peers to their ignorance, and even to their animosity. And he stuck to his guns. He wrote up a very fine anatomical and biometric study, comparing the Trinil femur to a thousand modern human femurs; he managed to extract the skull cap completely from its stone matrix and measured the extrapolated capacity; he did find the femur to be characteristic of an erect bipedal being, but still with numerous simian features; from the skull cap he estimated the total volume of the brain to be 855 cc, which would correspond, if this were an ape, to an animal weighing 230 kg, and, if it were a human, to an individual of 19 kg. He concluded, therefore, that "*Pithecanthropus erectus* is indeed unquestionably an intermediary between Man and the apes, a very venerable Ape Man, representing a stage in our phylogeny."

Conveying the ideas of the time, Marcellin Boule, on the other hand, wrote in 1913: "Why not suppose that *Pithecanthropus* represents an amplified, giant form of ape belonging to the gibbon group?" That is what Eugène Dubois in fact decided to do himself in 1935, but this time out of spite, to play down the importance of Ralph von Koenigswald's new discoveries of *Homo erectus* on Java. For although – like Philippe-Charles Schmerling in 1834 with the Engis Neanderthal, like Hermann Schaaffhausen in 1861 with the one from the Neander Valley, like George Busk in 1864 with the one from Gibraltar – Eugène Dubois had made a surprisingly pertinent study of his Indonesian fossils for a palaeontologist with no palaeoanthropological experience, he showed no quality of judgment regarding the new Javanese finds, nor indeed regarding those made in China from the 1920s, despite their being eminently comparable to the pioneering pieces. "That Peking Man is but another example of the Neanderthal race," he said of the Zhoukoudian fossils. As for those harvested in Java, he attributed them to the Neanderthals but also to Proto-Australians less than 10 000 years old and which he himself had in fact described during his first years on that island. "It is very regrettable," he wrote towards the end of his life, "that in interpreting the important human fossil discoveries in China and Java, Weidenreich, von Koenigswald and Weinert should have been misguided by preconceived ideas."

The discovery of Peking Man, also a *Homo erectus*, and, what is more, a Far Eastern *Homo erectus*, came immediately after that of Java Man in the story of the story of Man, "immediately" here meaning 30 years later, all the same.

The very first piece attributed to this new fossil Man, a tooth, was in fact found in 1921 by Otto Zdansky, a post-doctoral student of Carl Wiman, a palaeontologist at the University of Uppsala. It was on the advice of a Swedish mining prospector, Johan-Gunnar Andersson, a palaeontology fanatic, that Zdansky had opened a dig at Chou-Kou-Tien (today Zhoukoudian), 50 km from Peking (Beijing), and that is where he had found this molar which, he said, he recognized at once, but which he kept quiet about so as not to be in the spotlight ... It took the official visit to China of the crown prince of Sweden, five years later, for Zdansky to decide to mention his discovery – two teeth, in fact, the second having been recovered during the sifting of the gathered material – of "the most ancient type of human whose remains have been discovered in the layers of earth". Zdansky's discretion was justified; no sooner had the discovery been announced, than the anatomist of the Peking Union Medical College, a Canadian, Davidson Black, in effect seized it – there is no other word – published it himself in *Science* and *Nature*, and successfully applied for a generous grant from the Rockefeller Foundation to open a systematic two-year dig at Chou-Kou-Tien.

The site was opened during Easter 1927. Otto Zdansky, who was invited, declined the offer, and Carl Wiman, who had at least been consulted, sent a new post-doctoral student to China, a Swede named Birger Bohlin. Bohlin set to work and turned over 3 000 m^3 in six months; result: another tooth, three days before he stopped work! That was enough for Davidson Black, who could not stand it any longer, to announce the creation of a new genus *Sinanthropus pekinensis*, with which he associated Zdansky, and which he justified as follows: "The new specimen discovered shows, in the details of its morphology, a number of interesting and unique features, sufficient to justify the proposal of a new genus of hominid."

Davidson Black was in turn to encounter the scepticism and criticism we have come to expect in the course of this tale.

In the meantime the excavations continued to yield: half a mandible with three teeth in 1928, new isolated teeth in 1929 and, on 2 December 1929, the first skull, in fact discovered by Bohlin's most important Chinese collaborator, Pei Wenzhong. And during this time Davidson Black also succeeded in raising enough funds (still from the Rockefeller Foundation) to create a geology and palaeontology laboratory "of the Cainozoic", as he called it, which was to be the ancestor of today's IVPP, the Institute for Vertebrate Palaeontology and Palaeoanthropology.

But Davidson Black died suddenly in 1934, and a new team took over, with Pierre Teilhard de Chardin in the field and Franz Weidenreich as director of the laboratory. In 1937 they possessed 14 skulls, 11 mandibles, 147 teeth, 7 femurs, 2 humeri, the most beautiful collection of fossil humans ever put together. But for political reasons the excavations of this first period of 16 years were stopped; for the same political reasons, the precious fossils were packed away in cases and despatched to the shelter of the American Embassy, but the two cases were lost between Peking and their final destination (in fact probably between Peking and the port of Chingwantao where they were to be embarked at the end of 1941 on the SS President Harrison, which, it would also appear, did not show up). Nearly 30 years would pass before this site, so fertile, would be successfully reopened.

Despite the rather tepid welcome that was given to Peking Man, as had been the case with the other fossil humans whose discovery had preceded this one, many minds had gradually got used to the idea of a human form more ancient than Neanderthal and even more impressive on account of the thickness of his cranial bones, the flatness of his dome, the modest dimensions of his brain, the size of his teeth, and so forth. Java man had been castigated, but his existence and his status had, consciously or not, been admitted. Peking Man was thus to benefit from Eugène Dubois' battle and have less trouble being recognized, admitted and received.

A battle there would be nonetheless, but on another field, the cultural field – in the second degree, as it were. An important set of tools (essentially of quartz) had in effect been assembled at Chou-Kou-Tien; studied by Pierre Teilhard de Chardin and Pei Wenzhong, then by Father Henri Breuil, the set had, moreover, sometimes been accompanied by incontestable traces of fire. This was all that was needed to bring the worriers out of the woodwork: Marcellin Boule, for example, supposed that, contemporaneously with *Sinanthropus pekinensis*, there had lived another Man, a "true" Man, the beautiful Man we spoke of at the beginning of this chapter, and that it was obviously this other Man who had shaped the stones, lit the fire, and eaten the *Sinanthropus* associated with those remains!

But since he was never discovered, little by little it was agreed to grant *Homo erectus* the credit for manufacturing the objects and mastering fire.

Today a worrying suspicion still hangs over these Far Eastern *Homo erectus* specimens, with whom the *Homo erectus* specimens from Africa are now also associated. Some authors have thus suggested a *Homo habilis/Homo sapiens* filiation, short-circuiting *Homo erectus*; others, a parallel evolution: gracile australopithecines/*Homo habilis* and robust australopithecines/*Homo erectus*!

On the other hand, a tender gaze tends to be beamed towards the types collectively known as the European *Homo erectus*, who have been seen, in a convenient haze, parading as "Ante-Neanderthalians" or "Archanthropoids", or even "Palaeoanthropoids". It is certain that from the genetic drift onwards, this *"Homo"*, endowed with some apomorphic Neanderthal features, did not look altogether like his contemporaries from Africa or Asia, but it is also true that, whatever is covered by the terms *habilis, erectus* and *sapiens*, everywhere in the Old World *Homo habilis* little by little became more and more "erectus-like", *Homo erectus* little by little more "sapientized", while in Europe, the one who was there (*Homo habilis* or *Homo erectus*) little by little became more "neanderthalish".

It was a young doctor from Australia who had recently been appointed anatomy professor in Johannesburg, Raymond Dart, who had the good luck, in 1924, to hold in his hand the first skull of a new genus of hominid, illustrated today by some 5 000 pieces, and the extraordinary acumen to recognize it.

A student from his anatomy class (the only girl, as it happens), Josephine Salmons, had in fact brought Dart a fossil baboon skull found in breccia at Taung in the then Bechuanaland (now part of South Africa). Dart, very interested, alerted one of his colleagues from the Geology Department, Robert Burns Young, who hurriedly made a visit to the quarries of the Northern Lime Company one October day in 1924. The company in question was indeed in the process of exploiting the dolomitic limestone of the region for the production of cement and disposed of the breccia that filled the solution cavities. Young met the site foreman, A.E. Spiers, to request his authorization to pick up blocks of breccia, and great was his surprise to see on Spiers's desk, on a pile of mail and invoices, a little skull still embedded in breccia and split in two: one of the pieces represented a natural, almost disengaged endocranial cast or mould of the interior of the skull; the other, the back of a face for the most part hidden in the rock. Spiers, when questioned, declared that it was probably a Bushman skull with no more than a curiosity value and he happily gave it to Young who religiously wrapped it up and brought it back to his friend Dart. And Dart told how he received it just as he was preparing to attend a wedding ceremony, with his collar still unbuttoned, and the guests arriving. Exalted by what he saw, he immediately recognized, he said, human features on the mould of the skull interior but had to put it aside, for the bridegroom had arrived and was begging him to hurry!

Dart freed the skull from its breccia matrix, which was done by Christmas; by mid-January his descriptive and interpretative article was ready; in February it was published in *Nature*. In a curious mixture of excellent anatomical observations, on the one hand – the skull is small but the space between the sulcus lunatus and the parallel sulcus (two furrows found on the brains of apes and often in men, especially primitive men) is three times as large as on the brains of the great apes; the canine is small, the milk teeth are molarized, the enamel of all the teeth is thick, the situation of the foramen magnum (the hole through which the brain communicates with the spinal chord) suggests a skull positioned vertically on the spine and consequently an erect stance and bipedal locomotion – and incredibly fantastic declarations, on the other hand – the creature could appreciate colour, weight and shape; it knew the meaning of sounds and had already come a long way along the road to articulate language; it walked upright with the hands free to serve as organs of manipulation, aggression and defence – Raymond Dart, in turn, brought the australopithecine into the story of the story of Man.

The discovery of what was heralded as a new ancestor even less presentable than Java Man or Peking Man, themselves already less presentable than Neanderthal Man, could not but rouse criticism and invite rejection, all the more so since the information came from Africa, a continent from which nothing desirable was expected, and emanated from a young Australian doctor, a nobody, who had spent only three years in London (in an obscure post as an anatomy demonstrator at the University College).

The discovery did not therefore come as good news; what is more, there was something weird about it which deepened the frowns of London's anthropology circles: the australopithecine's head was back to front! More than ten years had in effect been spent debating the antiquity and ancestor status of *Eoanthropus dawsoni* before finally reaching consensus as to the reality of this long-awaited missing grandfather, which had the advantage of being not only European, first and foremost, but also, in spite of its still rather simian jaw, endowed with the reassuringly large head that was expected of an ancestor. And now here was this arrogant native landing on the scene of our origins, they said, with its little monkey face and its almost human set of teeth. Decidedly, Dart the exotic was going to be a thorn in the flesh of the all-powerful metropolitan scientific aristocracy.

Besides, the story of *Eoanthropus* is itself a strange one that has never been properly explained. Perhaps some light can be thrown on it by the way Sir Arthur Keith, an undisputed London authority on anatomy, entrenched himself in an opinion which ended up looking for all the world like a dogma. In 1888, a human skeleton and some stone tools were discovered at Galley Hill, to the east of London, in the gravel along the Thames. Anatomists identified the skeleton as very modern but geologists dated the gravel to the Upper Pleistocene, from which one may very reasonably conclude that a grave had been dug in the gravel. And yet, 25 years later, Arthur Keith decided to revisit this discovery, arriving at these surprising conclusions: the Galley Hill skeleton was indeed that of a *Homo sapiens*, he said; it came from a grave in the gravel but was nonetheless contemporary with the gravel. It is a typical case of that hope, so common in perfectly sincere – we hope – anthropologists, of discovering very ancient very modern grandparents for humanity, which one may call the Galley Hill syndrome, and which, later, would for example contaminate Louis Leakey in Kenya or Pierre-Paul Grassé in France.

Now, some years after the declarations of Arthur Keith (around 1912), an amateur geologist called Charles Dawson "discovered", in Sussex this time, near Piltdown Common, in gravel which he took to be from the Upper Pleistocene, fossils of vertebrates (elephant, mastodon), stone tools and strange hominid remains, a skull cap, four fragments of a parietal bone and first a mandible, then a lower canine, a fragment of a frontal bone and brow ridge, a molar and a fragment from the back of a skull.

Arthur Smith Woodward, in charge of palaeontology at the British Museum, was alerted, joined the search, reconstructed the skull and published the discovery with Dawson in the *Quarterly Journal* of the Geological Society in 1913. He then entrusted the description of the pieces to Grafton Elliot

Smith and, at his request, lent moulds of them to Arthur Keith; the former made a reconstitution of the capacity of the skull, arriving first at 1 070 cc then at 1 200 cc; the latter immediately at 1 500 cc. A mixture of sound observations – and excessive interpretations – ensued, emphasized, as if it were necessary, by a long-standing antagonism between the two Arthurs, Arthur Keith and Arthur Smith Woodward. Dawson died in 1916, Arthur Keith, Arthur Smith Woodward and Grafton Elliot Smith were knighted, and *Eoanthropus* limped his way along into treatises and monographs.

Thirty years were to pass – with tests carried out by Kenneth P. Oakley and M.F. Ashley Montagu of the British Museum to establish the contemporaneity of the bones by measuring their fluoride content (the content increases with the age of the fossil from the time it is first buried) – before the not very great antiquity, to say the least, of the human remains from Galley Hill (0,3% fluoride) and Piltdown (0,2%) was confirmed, compared with those from the Middle Pleistocene found at Swanscombe (2%) or with the Lower Pleistocene vertebrate fauna (from Piltdown or elsewhere) found at Piltdown (the previously cited elephant and mastodon fragments having a 3,1% fluoride content).

K.P. Oakley and C. Randall Hoskins were content at first, in a March 1950 article in *Nature*, to report this absence of contemporaneity between the fossil animals from Piltdown and the *Eoanthropus* remains; then, noticing that the teeth had been filed, that varnish had been applied to the skull fragments and that *Eoanthropus* finally turned out to be the association of a modern (present-day) Man's skull and an orang-utan's mandible, Joseph Walker, then at Oxford, went further and cried hoax. The greatest swindle in the history of palaeoanthropology was thus only revealed after 40 years; but although many eminent colleagues were suspected, including Arthur Keith and Pierre Teilhard de Chardin who had also dug at Piltdown, the culprit has still not been identified, after 90 years!

So when, in February 1925, the young anatomy professor from Johannesburg addressed the description of his little bomb to London, the official scientific community believed that the ancestor of Man was already known to them: indeed their *Eoanthropus* did have an ape's mandible – a minimal concession to Darwin – but he was above all blessed with a very large skull, containing a very large brain. The four experts to whom Dart's article was referred, Sir Arthur Keith, Sir Arthur Smith Woodward, Grafton Elliot Smith and Winfrid Laurence Henry Duckworth, all saturated with the conclusions of ten years of study on that ancestor, were in no hurry to let themselves be convinced by an almost diametrically opposed proposal; so they wrote pleasant enough reports – which enabled the article to be published – but all concluded that *Australopithecus*'s affinities lay rather with the great apes.

Generously, Sir Arthur Keith found a resemblance to the two great African apes: "one is inclined to place *Australopithecus* in the same group or sub-family as the chimpanzee and gorilla. It is an allied genus. It seems to be near akin to both." Arthur Smith Woodward, for his part, inclined rather to a connection with the chimpanzee: as far as he could judge from the photographs, he could see nothing in the orbits, nasal bones and canine teeth which was any closer to the human condition than the corresponding parts of the skull of a young chimpanzee of today.

Winfrid Duckworth said much the same, but favoured the comparison with the gorilla: "So far as the illustrations allow one to judge, the new form resembles the gorilla rather than the chimpanzee."

Then, in their various conferences, the tone, until then still somewhat cautious, became firmer and even derisive. Grafton Elliot Smith declared for example at a conference at the University College of London in 1925 that it was unmistakably an ape, a close relative of those still living in Africa, the chimpanzee and the gorilla, and that the creation of a new family was not justified, adding: "It is unfortunate that Dart had no access to skulls of infant chimpanzees, gorillas or orang-utans." And the same year, Sir Arthur Keith went one further: Dart had found an extinct relative of the chimpanzee and the gorilla, he said, but with some human features; his discovery threw light on the history of the anthropoid apes but not on that of Man. Robert Broom, a Scots doctor and palaeontologist living in South Africa, who completely supported Dart's interpretation, wrote while on a trip to London in 1928 that he had been to Europe and had been amazed to see that the matter was closed: the Taung skull was that of a chimpanzee and that was that.

Raymond Dart had moreover greatly irritated the happy few by confiding in the press without respecting the priority of their verdict; before his paper was published in *Nature*, many newspapers had, in a flurry of headlines, announced the Taung discovery: "Birth of Mankind", "Missing link that could speak". The South African pavilion at the Empire Exhibition in 1925, in Wembley, displayed the endocast and some attempted reconstructions of the young Ape-Man, while the authorities in the profession did not yet have access to the least facsimile. It was through the glass window of the exhibit case that Sir Arthur Keith was to make the first three-dimensional observation: "The skull," he declared after his visit, "is that of a young anthropoid ape – one which was in its fourth year of growth, a child – and showing so many points of affinity with the two living African anthropoids, the gorilla and chimpanzee, that there cannot be a moment's hesitation in placing the fossil form in this living group."

This did not stop Raymond Dart from proceeding with his detailed description of his specimen; he completed it in 1930, sent the monograph to his publisher and took himself off to London in 1931 with the fossil in his pocket! He was invited to speak at the Royal Zoological Society, but his presentation failed absolutely to convince anybody. He wrote:

> I stood in that austere and chilly room, my heart pounding with the hope that the expressions of polite attention on the four score faces before me might change to vivid interest as I spoke. I realized that my offering was an anti-climax but with undiminished optimism launched into my story. What a pitiful difference between this fumbling account and Elliot Smith's skilful demonstration! [He had just returned from China and was speaking about the Peking Man fossils, which he had inspected.] I had no plaster casts to pass around, no lanternslides to throw on the screen to emphasize my points. I could only stand there with the tiny skull in my hand, telling the audience what I saw as I looked at it – all of which had been previously published, with illustrations.

Raymond Dart once told me how, during this first visit to London with the original, the "missing link" had nearly gone missing. He had secured an appointment with Sir Arthur Keith, had wrapped the two pieces of the little skull in newspaper, had called a taxi and, accompanied by his wife (his first wife, he pointed out with a wicked wink), had set off for the appointed place. Arriving at their destination, the Dart couple alighted from the taxi, but without the precious package. It took Scotland Yard some time to retrieve it from the kitchen table of the taxi driver, who had gone home, had found the weird forgotten object and had no idea what to do with it!

The confusion between great apes and prehumans has of course cropped up on several occasions since then, plesiomorphic features being taken for apomorphic features. How many authors have referred to Lucy as a panid (from the genus *Pan* – chimpanzee), classifying the inherited features from the common ancestors of prehumans and great apes, so useful in reconstituting the identikit of those ancestors, among the derived features which form part of the definition of the great apes.

At the end of the 1960s in France, a certain number of personalities, whose authority in the anthropological disciplines was undisputed, and rightly so, snubbed the australopithecine-human filiation. Admittedly French palaeoanthropologists had never had access to original australopithecine material before our discovery (Arambourg and Coppens), from 1967 onwards, of australopithecines in Ethiopia. Very well established and highly reputed, however, was the French tradition of the study of *old Homo sapiens* (Cro-Magnon, Chancelade, Grimaldi, Combe-Capelle), Neanderthals (La Chapelle-aux-Saints, La Ferrassie, La Quina, Le Pech de l'Azé), old Neanderthals (Montmaurin) or true *Homo erectus* (Rabat, Sdi Abderrahmane, Ternifine).

But, out of the information arriving from the other side of the dark continent, people like Jean Piveteau or Henri-Victor Vallois and, a fortiori, before them, Marcellin Boule or Paul Rivet had only been able to form incomplete opinions emanating from a few readings, photographs and casts. My application for the deputy directorship of the Musée de l'Homme in 1969 was to be adjudicated at a meeting of the professors of the Muséum National d'Histoire Naturelle. One of them, Henri-Victor Vallois, a former professor of anthropology at the Muséum and former director of the Musée de l'Homme, remarked, I was told later, that in spite of my qualities, which he recognized, I had a handicap – a serious handicap: I studied apes (the australopithecines), and that was hardly the best preparation for taking on that responsibility in an institution devoted to the study of Man.

Today, the prehuman (australopithecines)/human filiation is more universally accepted, but for some authors a certain nostalgia undoubtedly persists for ancestors who would turn out to be beautiful, old, real humans; the gentle australopithecines, complete hominids, who were the possible craftsmen of the first deliberately conceived tools, are for example still called bipedal apes by some (Jean Chaline).

CHAPTER 4

AUTOBIOGRAPHY
The Story of My Story of Man

זְכֹר יְמוֹת עוֹלָם בִּינוּ שְׁנוֹת דּוֹר־וָדוֹר
שְׁאַל אָבִיךָ וְיַגֵּדְךָ זְקֵנֶיךָ וְיֹאמְרוּ לָךְ׃

*Remember the days of old,
consider the years of many generations;
ask thy father, and he will show thee;
thy elders, and they will tell thee.*
The Pentateuch, Deuteronomy,
Chapter XXXII, verse 7

The first 20 years of my professional life leave me with the impression that they were the bridge between a long, small-scale, home-baked and very national past, which did of course also mould me, and a later situation where production, evaluation, communication and competition had acquired a new, more global dimension that I, in my own small way, actually helped to bring about.

In this fourth, somewhat autobiographical chapter, therefore, I shall attempt to describe the carefree serenity of the "pioneering years", when half a dozen people endowed with a keen sense of observation sufficed, side by side, to form a large laboratory; the joyful passion of the "crazy years" which, because they were richer, saw the organization of the grand expeditions and their generous teams of dozens of specialists who were immediately available and operational; and the austere rigour of the "heritage years" which followed, when specialization remained but was enhanced by a greatly improved ability to quantify in order to convince.

But, the reader may rest assured, enthusiasm was no more lacking in the pioneers than it was in their heirs; it is simply that the world has evolved; its economy has changed, and so have science and its methods.

The world has clearly opened up to such an extent that it would be practically inconceivable today for a team to undergo peer assessment only by compatriots – even the word "compatriot" has fallen into disuse. The economy, on the other hand, has fluctuated to the point where, after a long period of discretion in its ivory tower, science could for some 20 years enjoy the recruitment, the funding and the popularization it needed to carry out operations that were unthinkable and unheard of until then, as well as progressively to transform its mentalities; and then, while the structures and practices of scientific information and communication have remained intact, circumstances have recently cut back our manpower and resources once more. On another level, there has been an unprecedented proliferation, improvement and diversification of all kinds of techniques transcending the boundaries of many disciplines including our own, to such an extent that we have had to familiarize ourselves with the practices of other disciplines, where mathematical calculation plays an important part.

In other words, we work better and better, communicate our findings and our interpretations of them ever more effectively, but the development of our human resources and our materials and equipment remain dependent on the goodwill of politicians and their decisions.

I was born in Brittany and spent my first seven years there, then, after an interruption of three years in the vicinity of Paris, 12 more years in Brittany divided into seven and five, seven at high school and five at university. None of this would be very important if my interest in the past had not manifested itself right from the first slice of this division and had not gone on developing steadily in confidence and vigour from then on. So it was in Brittany, and no doubt under the influence of its landscapes dotted with huge prehistoric stone monuments (called megaliths), that this attraction, soon tinted with passion, insidiously took root in my head. And if, in the third and fourth slices of the same division, I sat for the particular baccalaureate then called the *Baccalauréat de sciences expérimentales* and later for the Bachelor's degree then called the *Licence de sciences naturelles*, it was to prepare myself to live a professional life devoted to that vocation.

What is good about university, when one is research oriented, is that it allows immediate access to laboratories, to their teams and to their facilities, and any faculty gives one access to all the others; from the Science Faculty of the University of Rennes, which was my alma mater, I thus joined the Science Faculty in Paris (then the Sorbonne), but also the Faculties of Arts and Medicine in both Rennes and Paris, and a certain number of institutes with public or private university status: the Rennes Institute of Geology, the Paris Institute of Art and Archaeology, and the Monaco Institute of Palaeontology (in Paris). Now in the disciplines practised in those establishments and which revolved around my chosen area, the names of certain pioneers dominated the scene, such as Giot, Merlat, Piveteau, Demargne, Vaufrey, Vallois, and so on. And so it happened that I took part in the palaeontological, prehistoric and historic excavations of some of them, broke stones or dissected with others, and soon enrolled for doctoral studies at the Sorbonne, in the Laboratory of Vertebrate and Human Palaeontology, which also recruited me almost immediately as a researcher, thus giving me entry to the CNRS (National Centre for Scientific Research). That was in October 1956.

Some months later, the new management of the Palaeontology Laboratory of the Muséum National d'Histoire Naturelle appealed to some of the young researchers from the Sorbonne's Palaeontology Laboratory to come over with some new ideas and energy to the rather deserted Jardin des Plantes and the extraordinary collections that it housed. I was one of them.

In this way, during the last few years of the 1950s, I got to know, from the inside, the research teams I had just joined as well as those I was already part of, those of the Palaeontology Laboratory of the place Valhubert – very soon to become the Institute of Palaeontology of the Muséum – and at the same time those of the semi-basement laboratory at 54 rue Saint-Jacques, still very student oriented, which I had not really left. On the Jardin des Plantes side, Arambourg (newly retired), Lehman (newly arrived), Lavocat, Hoffstetter, Roger and Ginsburg were the masters or elders, always models, in a sense; on the Sorbonne side, there were Piveteau (a lot), Devillers (a little), Dechaseaux, Guth and Genet-Varcin (very little) who played the same roles there. Since elsewhere in Paris, at the same time, the following also still prevailed: Breuil, Lantier, Rivet, Vallois, Grassé, Termier and Termier, Anthony (Jean), Vaufrey and Leroi-Gourhan in the tradition of Boule, Piette, Teilhard, Déchelette, Anthony (Raoul) and Joleaud, my roots like those of my friends and colleagues of the time went deep into a humus of palaeontology and prehistory in the purest tradition of the beginning of the twentieth century and indeed even of the end of the nineteenth century.

Moreover, as I knew Father Breuil (known to fame as l'abbé Breuil) very well, and he had known the Fathers Bouysonnie, Denis Peyrony and Dr Henri Martin, the respective discoverers of the Neanderthal Men from La Chapelle-aux-Saints, La Ferrassie and La Quina; and as I was good friends with Ralph von Koenigswald, the discoverer (at Sangiran) of an impressive collection of Java Men, who himself had known Eugène Dubois, the discoverer (at Trinil) of the first of those Men; and as I was a friend of Raymond Dart, who had described the first australopithecines found in Taung – Raymond Dart who had naturally known Dr Robert Broom and John Robinson very well (I also knew John Robinson), the discoverers (at Sterkfontein) of the next australopithecines – clearly the period of my entry into professional life – 1950–1956 – enabled me to become closely acquainted with the first discoverers of certain of the prehistorical hominids (australopithecines), or with those who had known the first discoverers of certain others (pithecanthropines, sinanthropines), or still others who had known those who had known the first discoverers of still others (Neanderthals).

In other words, I had the good fortune to experience the last of the pioneering years of the very young sciences of prehistory and palaeoanthropology, which I was starting to serve.

Of all those pioneering personalities, Camille Arambourg was certainly the one I was closest to. Camille Arambourg had just retired as professor of palaeontology from the Muséum National d'Histoire Naturelle when I met him in his own laboratory, the one he had inherited from Marcellin Boule in 1936 and which he had directed for 20 years before handing over then to Jean-Pierre Lehman. In order to be an "ideal" professor at the Muséum, considering the multiple tasks expected of the incumbent by the institution, one had to be all at once a teacher, a curator and a researcher; since it is exceptional to possess those three talents equally to the highest degree or at any rate to have the time to exercise them all, each professor is generally one or two of the three, but rarely all three of them. One could say, without passing a value judgment, which I would not permit myself moreover, that Camille Arambourg was more of a researcher and Jean-Pierre Lehman, after his appointment at the Muséum, more of a teacher and curator.

And so it was that Camille Arambourg and I often bumped into each other in front of the drawers of the old wooden cabinets of the great palaeontology gallery, overcrowded with exhibits, commenting on some or other dental morphology or the abnormal development of this crest or that groove. And then I left to go prospecting in Africa and, after a difficult but fruitful first expedition, returned there. I had the impression that for Camille Arambourg, who, at more than 75 years old, had not really had any pupils nor, presumably, felt the need for any, this was, suddenly, the possibility of securing the collaboration of a young scientist who, like himself, was in love both with palaeontology and the vast African horizons that he had frequented so often. Our relationship indeed grew closer; he would call me into his office more and more often, or to the little laboratory next door, to show me, for my information or to ask my opinion, a fossil that he was busy describing or that he had just received. And one fine day, Camille Arambourg, who had just turned 80, summoned me to declare, briefly, and with intense but contained emotion, that he had decided to make me his scientific heir. A few months later we were digging together in Algeria; the following year saw us prospecting together in distant corners of Ethiopia, the Sudan and Kenya. Three years later he died in Paris, after the two of us had just put the final touches to the programme of the fourth campaign in East Africa. He was nearly 85.

Although of the generation of his grandchildren rather than of his children, I can say that I was to some degree modelled by the example that this grand old man gave me. Of course he was only one scientific personality who cannot alone illustrate by his professional activity, conducted between 1912 and 1969, what I have called the pioneering years, but he was, all the same, one of the great figures of that period who are no longer with us.

Camille Arambourg (Fig. 5) was an agronomical engineer with a passion for geology and palaeontology. He did in effect first teach geology at the Agricultural Institute in Algiers and at the National Institute of Agronomy in Paris before joining the Muséum. He worked alone, technically assisted by a few loyal collaborators who prepared his fossils and his manuscripts, producing the casts and photographs of the former, or researching the bibliographies and composing the plates for the latter. He would receive, with all due courtesy and attention, the half dozen or so foreign researchers who would come every year to see particular pieces in the collections, but he was not keen to take the initiative in organizing seminars or exhibitions in order to attract, assemble or retain them. Communication existed, naturally, between palaeontologists from all countries, but was not a priority because the need to develop it was not felt. We had once received a favourable and prompt response from overseas to a request that he and I had formulated but whose point now escapes me, and I can still hear Camille Arambourg saying to me: "We shan't reply right away; it always makes a better impression to let some time pass, doesn't it?" We were far from the days of e-mail! However, when in Paris, he was at the laboratory from morning until evening, and very regular in his office hours, which were interrupted only to go home to the rue Lagarde for lunch.

Camille Arambourg would often go out into the field to prospect, dig, assess a site or a finding; he only exceptionally went to other laboratories, other museums, seminars and conferences. In the field, with only rare exceptions – his work site at Ternifine in Algeria, and ours in the Lower Omo Valley, in Abyssinia as he called it – he also kept to himself. When called by an informant, a geologist, an engineer or administrator from the Ministry of Roads, he would go with the person to the relevant place, locally recruit a few labourers if the deposit looked promising, and would return to the Muséum with his harvest which he would have his assistants prepare and which he would then study entirely on his

Fig. 5 – Camille Arambourg (right) and Louis Leakey, photographed in the field by the author, in the Lower Omo Valley in Ethiopia (1967).

own. And then would follow adequate and generous publications, announcements, partial descriptions, monographs, syntheses, broader overall reflections, possible general revisions – in a word, a sound and intelligent exploitation of the newly collated material, until he would go out again to search for more.

Jean Piveteau, a contemporary of Camille Arambourg, was, for his part, an academic; an important part of his activity was always geared to his teaching, which was, moreover, renowned, and to the training of aspirant teacher-researchers. He was also a brilliant researcher himself, with such broad insights that after completing a treatise with Marcellin Boule, he was the sole editor, admittedly with the precious collaboration of Colette Dechaseaux, of a monumental treatise in six volumes, which he completed in 1957 with the volume devoted to Man (written entirely by himself). As far as fieldwork was concerned, on the other hand, apart from a few expeditions reluctantly undertaken, including a particularly fruitful one to Madagascar, he did not enjoy going to hunt for fossils himself; he was a scientist – a great scientist – but an indoor scientist; he once told me, at the time when I was working in the Sahara in Chad, that he had caught a glimpse of the Sahara one day in Southern Tunisia, and that "it was frightening" – with much humour, naturally. As for the savannah, whose gentle beauty and fragrances I had been extolling, he had visited it only once, during a private trip to Senegal, and had described the scenery to me in these terms, with the same humour: "Oh! you know, a baobab here and there ..."

The intermediary generation, and for me still a pioneering one, between the great outdoor palaeontologists (Camille Arambourg) and the great indoor palaeontologists (Jean Piveteau) had pursued the tradition of that solitary research and those solitary expeditions; Robert Hoffstetter in Latin America and René Lavocat in Africa, for example, had acted no differently from Camille Arambourg.

Only Jean-Pierre Lehman, possibly on account both of his university training – he had been Jean Piveteau's assistant and had remained very close to him – and his new responsibility at the Muséum, had retained and striven to perform the multiple duties of teaching, research training and team building, setting up temporary exhibitions and redesigning the techniques of permanent museum displays, organizing large exhibitions (Spitzberg) and international conferences. Perhaps one could retain his example, together with that of Louis Leakey – in this classification which is obviously altogether arbitrary and for which the author assumes sole responsibility – as a pivotal model between that of the pioneering years and that of the crazy years.

Louis Leakey (Fig. 5) was another of my masters; although very British – I know what I am talking about – he had had the generosity to take an active interest in my early research, to the point of supporting me by addressing letters to my tutelary institution, the CNRS (the N – for "national" – in this acronym represents France), introducing me to American foundations, having me invited to their conferences, securing funding from these foundations, and so forth. He invited me to join him in the field in Tanzania in 1961, but I could only honour his call in 1963, as the 6th regiment of the sappers in Angers to which I had been assigned for eighteen months had not considered this invitation important enough to warrant special leave (at the same time, a friend, in the same situation and the same barracks, had been granted the exception, which was a cultural one par excellence, to compete in the Prix de Rome for painting, which he in fact won). Louis Leakey did not in fact work very differently from Camille Arambourg; even though he was indisputably more of a builder than the latter, since he was forever adding to the buildings of the Coryndon Museum that he directed in Nairobi, his prospecting or digging teams in the field were often very modest and local.

But there was a turning point in Louis Leakey's life and behaviour: the discovery in 1959 at Olduvai of the australopithecine skull called "Nutcracker", its "absolute" dating in 1961, and the ensuing clever scoop by the American, that is to say global media – one must be realistic – of the double event and of the immense public interest that it generated.

"It's true that that is what sparked it off," agreed Mary Leakey, Louis' wife, one day after hearing me tell her own story, as though she suddenly realized that it was that dated discovery, by its global impact and its financial repercussions, which had changed the scale of the programmes and research operations undertaken.

In fact Louis Leakey had first prospected around Olduvai in 1931, 28 years before. It was only fair that his patience should have been rewarded. There had indeed been some earlier discoveries, a first hominid before Leakey, but whose age had rightly been criticized (it was a man dating back about 17 000 years – the time of Lascaux – buried in layers of approximately one million years old!), some bones of a *Homo erectus* skull in 1935, an australopithecine tooth in 1955, and tons of remains of vertebrates and stone tools throughout that time, but not enough to stir the conscience of the world and get it to subsidize the work sites, not even enough for the scientific community, more inert than one sometimes imagines, to turn its spotlight towards East Africa. It must also be said that Louis Leakey, always very busy with many other sites, had not maintained a permanent work-site at Olduvai between 1931 and 1959; he would go there from time to time and would advance, from dig to dig, towards a better understanding of the history of these layers and their contents.

A few dozen kilometres from Olduvai, on the shore of Lake Garusi, a German expedition led by the palaeontologist Ludwig Kohl-Larsen had in fact already, on the eve of the Second World War, found some remains of genuine very old hominids in an eloquent context of other fauna, but the discretion of their publication and the confusion of their classification – *Meganthropus* and then *Praeanthropus* – had not marked these finds as being of much importance.

Louis Leakey himself – but he died without knowing it – had in 1935 found an australopithecine (lower left) canine in the same layers, which today are generally called the Laetoli layers, but he had not recognized it; the credit must go to the brilliant and mischievous Timothy White for realizing this.

It therefore took the discovery of this superb skull with its complete upper set of teeth for palaeontologists to begin to take note of that part of the world in their reflections on the origin of Man. As Louis Leakey had the further merit of thinking of finding a way, in collaboration with physicists, of accurately dating his discovery, and as this attempt resulted in the extraordinary figure, for the period, of 1 750 000 years, obtained for a volcanic level just above the level which had contained the skull (a dating obtained by measuring the disintegration – from the time of the volcanic eruption responsible for the explored level – of the radioactive potassium it contained into radioactive argon), scientists suddenly only had eyes and ears for this province of the tropics.

There was in fact at that time one of those backlashes of scepticism and rejection so familiar in the sciences, and particularly in palaeoanthropology, as we saw in the previous chapter. In Paris, in prehistory and palaeontology circles, they trusted the two physicists, Evernden and Curtis, who had performed the dating – one is always impressed by what one does not understand – but they said that the skull had been found on the surface and that the proclaimed dating did not concern it. In fact, those who knew what they were doing had indeed understood that this was a major discovery and that the age attributed to it had to be taken into account. In other words, three obvious facts suddenly imposed themselves on the reflections of palaeontologists: the australopithecine had lived in East Africa; not only was it present there but it seemed to have been there for a very long time; there was moreover in East Africa a volume of accumulated sediments in the Rift Valley which represented an immense potential field of investigation.

This literally triggered a scramble for bones (an "old rush" rather than a "gold rush") such as had never before been witnessed in the profession since its origins, a rush which would have been qualified as "historical" if we had been on television! This was the starting point of what I shall call here the "crazy years", which were to last for all of two decades.

Counting only the important expeditions which were then to head for East Africa and conduct several campaigns each, three, four, five, ten, sometimes twenty, it is amusing to note that after hardly one year of inertia – or preparation – following the proclamation of the dating of the Olduvai skull, there would be practically one large new expedition per year for about ten years:

1963, beginning of the Harvard expedition (B. Patterson) south-west of Lake Turkana in Kenya; 1964, the Nairobi Museum expedition (R. Leakey, G. Isaac) at Lake Natron in Tanzania; 1965, beginning of the French CNRS expedition (J. Chavaillon) in the upper valley of the River Awash in Ethiopia, which was to last for some 15 years without ever really closing; 1966, beginning of the London Bedford College expedition (B. King, W. Bishop) in the Lake Baringo basin in Kenya; 1967, beginning of the international expedition (CNRS, University of Chicago, Nairobi Museum) on the Omo in Ethiopia (C. Arambourg, Y. Coppens, L. Leakey, R. Leakey, F.C. Howell), which was to last for ten years; 1968, beginning of the Nairobi Museum expedition (R. Leakey, G. Isaac) east of Lake Turkana, in Kenya, extending to the west of the lake (today under the authority of M. Leakey and A. Walker) 15 years later; 1972, beginning of the international expedition (CNRS, Musée de l'Homme, Cleveland Museum) in the Afar in Ethiopia (Y. Coppens, D. Johanson, M. Taieb), which would last for six years; 1974, beginning of the American so-called Ethiopian Rift Valley expedition (J. Kalb); 1979, beginning of the University of California expedition (T. White, J.D. Clark) in the Middle Valley of the River Awash in Ethiopia, interrupted three years later for local political reasons.

To these eight large-scale operations must of course be added the huge involvement of the Leakeys, father and mother (Louis and Mary), at Olduvai, then at Laetoli in Tanzania, and a certain number of lighter operations (for example those of the Savages, or of M. Pickford, or of D. Pilbeam in Kenya); all told, it can be estimated that during those 20 years (1960–1980) at least 500 people including 200 scientists must have explored some 2 000 km of the rift (Eastern Rift Valley) running from south to north, across Tanzania, Kenya and Ethiopia. The results matched the investment in manpower and equipment – one must face facts: 200 000 fossil vertebrates including 2 000 hominid remains belonging to the last 10 million years (but especially to the last four), tens of thousands of stone implements, but also an impressive harvest of fossil woods, pollens, shells, diatoms and ostracods, etc.

These results also clearly concern the incredible wealth of knowledge that has been extracted from them – Volume 6 of Jean Piveteau's treatise published in 1957 was obsolete in 1977; this knowledge included the precise datings that had been calculated, which were to make it possible for the first time to realize how old humanity was and to calibrate its history, and which would also transform the perception of the time dimension in the mentalities of an ever-increasing number of people. Also to be affected by these results was the style of research in the field, which quickly became much more multidisciplinary, more collective and hence more complete. And finally the perception of the duty to keep the public informed and the means to achieve this would undergo a profound transformation.

If, as we have seen, the 1 750 000-year dating had thrown out the palaeoanthropologists' calendars – no hominid fossil reached the million-year mark in the teachings of the Sorbonne at the end of the 1950s – the 3 000 000-year dating for Lucy or the 3 500 000-year dating for the Laetoli footprints no longer caused a stir at the end of the 1970s. The potassium-argon measure proposed by (Louis) Leakey, Evernden and Curtis, and applicable to most volcanic levels, was afterwards used almost systematically whenever possible and was soon enriched by a number of other so-called absolute methods based on the same principle of the disintegration of one radioactive element into another (uranium/thorium, rubidium/strontium or argon 40/argon 39).

Having co-directed, as mentioned, the international Omo expedition during the ten years of its operation in the field (alone with Francis Clark Howell after the death of Camille Arambourg in 1969, thus from 1970 to 1976), I had the pleasure of participating in drawing up a stratigraphic sequence of reference for the entirety of the East African sites of the last 4 million years. The Plio-Pleistocene deposits of the Lower Omo Valley in effect represent a column of more than 1 000 metres (1 100 metres) of material, sedimented between a little more than 4 million years ago and a little less than 1 million years ago. Now this column consists of accumulations of detritus – sands, clays, silt – and interstratifications of lava or volcanic ashes. In this 1 100-metre slice of deposits there are 106 volcanic levels, essentially of tuff, and most of them potentially datable.

As these terrains were moreover full of vertebrate remains enabling the construction of the finest biochronology in existence (of which I was the architect) and as the sediments had been the focus of a great many samplings in order to establish the palaeomagnetic polarity (as with the absolute dating processes, it was Frank Brown who had overseen these), the Omo deposits, thanks to their three scales – biostratigraphic, magnetostratigraphic and radiographic – all three of them exceptional, easily became the standard by which all contemporary deposits in East Africa and even Africa as a whole could be dated, by comparing methods.

The grand period of the crazy years, in other words the period of the great multidisciplinary, multinational teams who had come to prospect, dig, sift through and plough open the East African cradle, ended around the 1980s. The incredible commotion that prevailed in sanctuaries such as the National Museum of Kenya in Nairobi or, less intensely, the National Museum of Addis-Ababa or that of Dar-es-Salaam, slowed down noticeably to make way for a more manageable current of activities. The reason for this seems to have been twofold, natural and political; natural because, after 10 years, or even 15 or 20 years of uninterrupted activity in the field, the accumulated material required a period of analysis and the result of that analysis in turn called for a period of reflection, even if it was only to head for the field once more with new questions; political, especially in Ethiopia, where the situation no longer enabled researchers to work in safety in the field and where the field was soon completely out of bounds – research permits were refused between 1982 and 1990; both the Omo and Afar (Hadar and the sites in the Middle Valley of the Awash) closed their doors in Ethiopia, but so did Olduvai and Laetoli, in Tanzania, while in Kenya Richard Leakey crossed Lake Turkana, thus switching from intensive exploitation in the east to extensive prospecting in the west, and progressively handing over to his wife Meave.

I have called these years by another name than the preceding ones because they really did represent another period – shorter stays in the field, smaller teams, more numerous and varied laboratory analyses – a turning point in relation to the effervescence of the previous period, and the name I have chosen is the "heritage years" because they really did inherit the hundreds of tons of material gathered by us (50 tons from the harvest of my Omo expedition alone).

We, the Rift companions – in alphabetical order Chavaillon, Coppens, Howell, Leakey (mother and son), Taieb and our collaborators – were also in effect the first beneficiaries of this collected material, from which were to flow forth first of all monographs on Olduvai (Mary Leakey, Phillip Tobias), the east of Lake Turkana (Richard Leakey, John Harris, Bernard Wood), the Omo (Jean de Heinzelin and three volumes under the direction of Yves Coppens and Francis Clark Howell), Hadar (Donald Johanson and collaborators), and so on.

A great number of Ph.D. theses, some of them published, were also to result from this material, each studying a part of the hominid fragments collected or a part of their taxonomy, extending the study comparatively to other parts of the skeleton, other regions where such remains were found, other periods in which they had been found or other branches of their phylogeny. This group of young doctors has of course considerably swelled the ranks of the palaeontologists, palaeoanthropologists, prehistorians and archaeologists who little by little have become professional and operational and, since these young people have, like those before, become not so young, they too have directed doctoral theses and produced younger researchers who in turn have further increased the "palaeo" population that some of the older ones have not yet left ...

Laboratory work is what occupied them all first (while occupying us at the same time) and the very often excessive debates on datings, filiations, functional or behavioural interpretations followed.

At the heart of living science, while the chorus of performers grows and competition along with it, some false notes have obviously been heard from certain minds imagining that to diametrically oppose well-established opinions could constitute good strategy; and so one reads from time to time that the orang-utan is the closest cousin to Man, the australopithecine the ancestor of the chimpanzee, that erect stance and bipedalism were born in the forest or could just as well have been born there, and that in any event, it was the head which made the feet and not vice versa. And many scientific journals, even among the more respected ones, have enjoyed this. It is good not to accept the accepted and to keep one's mind open to the unexpected; but the deliberately contrived unexpected and a lack of judgment have never been the source of great ideas.

On the whole we also experienced wonderful analyses inspired by new techniques and leading to astonishing results; this period knew how to put mathematics to good use, sometimes learned elements of engineering, formed alliances with biochemistry, molecular biology, X-ray technology with its slicing potential, and naturally solicited all the possible dating techniques.

The interpretation of the signs of microscopic wear of the teeth had for a long time been utilized to identify diet. The way in which teeth are worn reflect both natural milieu and cultural behaviour; but, thanks to the extraordinary progress of the chemical industry and computer science, the new analyses went a great deal further in the quantification of these signs of wear by using various newly developed resins to take the imprints, allowing a much greater resolution, geometrically characterizing the said imprints through three-dimensional profilometrics, and finally, statistically processing the images thus obtained. The use of the microstructure of dental enamel (lines of Retzius, growth increments of Hunter-Schreger) had also been known for a long time in measuring the growth period of the crowns and had led to conclusions of systematic and environmental value; but, thanks to the progress in microscopy and, once again, in computer science, the new counts of the growth increments, their number or their rhythm – according to the tooth or part of the tooth under observation, naturally using all the skills of statistics – have also gone much further.

We thus learn for example that the "(H)Omo event", in other words the great desiccation 2,5 million years ago that I had brought to light 20 years earlier in observing the instability of the distribution of the fauna and its tendency to transformation, is now also read in the inside of the teeth by Fernando Ramirez Rozzi. We also learn that the flesh diet of the Neanderthals was prevalent in the cold periods of glaciation while vegetarian fare dominated between glaciations – "show me how you've worn your teeth down and I'll tell you in what period and conditions you lived" could have been Catherine Ussunet-Zarrouk's conclusion to her thesis, as she easily placed the dental remains from Krapina in Croatia, until then all higgledy-piggledy, in their original stratigraphic order.

The explorations into the interior of bones are no less fascinating for their new possibilities and the interpretations that they allow; the X-ray and the associated scanning technique which has already become a classical tool have resulted in superb images of the intimate secrets of our skeleton, enabling quantified measurements of bone mass (thickness, composition, density, sponginess) photographs of the internal supporting structures of bones (high density trabecular structures naturally aligned in the main directions of the biomechanical needs) and even three-dimensional reconstitutions of crushed or missing parts, or artificial assemblies of parts belonging to several individuals.

We thus learn from Renée Angelina Garcia that physical overactivity can be the origin of the development of cortical bone, but that over-development of the cranial vault of *Homo erectus*, essentially that of the diploe, is probably of genetic origin (perhaps hormonal); we learn too from Valérie Galichon that the criss-cross manner in which the trabecular features are organized within the bones of the pelvis clearly indicates the type of locomotion; finally we have seen Marc Braun projecting the virtually reconstituted skulls from Steinheim and Casablanca, filmed from every angle with every suture, although the first is in reality deformed by the pressure of the sediments and the second is represented only by the little frontal bone of one skull and the maxilla of another.

We had known for a long time what bones were made of, but thinking, in rather facile fashion, that old meant fossilized and that fossilization meant total and inexorable replacement, molecule by molecule, of all organic matter by mineral matter, we had not gone very far into the interior of fossil bones to see what was left. Palaeohistology was the first to do so, surprising us by its reading of growth, irrigation and thermic regulation; and then bio-geochemistry went deep into the isotopic content of the proteins preserved in the bones, the memory of the isotopic content of the consumed organic elements; molecular palaeogenetics, believing in miracles – and rightly so! – is currently attempting to identify the structures of ancient DNA.

Marc Fizet has taught us that the Neanderthals from

Marillac in Charente, about 50 000 years old, loved meat as much as wolves did and, when they had the choice, preferred reindeer steaks to bison steaks (established from the carbon 13 and ozone 15 in the collagen of the bones); as for Mathias Krings, he has just confirmed, by the molecular route, Neanderthal's singularity (or at any rate that of the Neanderthal Man from the Neander Valley), going perhaps as far as his specificity – the interfertility barrier between him and the other contemporary Man, Cro-Magnon.

Borrowing from mathematics its techniques of form analysis – the concept of conformation – has made it possible to account for a certain number of evolutionary changes that measurements, indices, surfaces, volumes, trigonometrical relations previously revealed imperfectly or not at all; borrowing from engineering its methods of movement analysis has made it possible to locate topologically and follow by means of synchronous photographs the volume movements of studied locomotions, and to reconstitute the traces.

Xavier Penin, borrowing from mathematics, for example demonstrated the carriage of the head and the general posture of the australopithecines (never humbly stooped as they are shown in classical imagery); Christine Tardieu, getting to grips with engineering, showed how, contrary to traditional representations, there had not first been quadrupedalism and then bipedalism like ours, but an evolution of bipedal locomotions and sometimes even a coexistence, confirmed by Anne-Marie Bacon, of several such locomotions.

But these heritage generations did also undertake fieldwork themselves, in smaller teams than in the crazy years, for financial reasons, making up for the absence of specialists on site by their laboratory consultations.

During these last two decades, prospecting has thus indeed been undertaken in vast regions which had never had visits from palaeontologists before or had only received them very long ago: Uganda, Botswana, Angola, Namibia, Malawi, Eritrea, Djibouti, to name only examples from Africa between the tropics. This has resulted in some very important discoveries of new great apes from the Miocene (Namibia), new prehumans (Eritrea, Malawi), new very old humans (Malawi, Uganda) or not so old (Djibouti, Namibia), and the confirmation (Uganda) of the existence of the climatic crisis dating back 8 000 000 years ago and which I have called the "East Side Story". New work-sites have of course been opened in the highly productive countries of the 1960s and 1970s, Ethiopia, Kenya, South Africa; it was these operations which enabled the discovery of *Ardipithecus ramidus* (Ethiopia), *Australopithecus bahrelghazali* (Chad), that of new *Australopithecus afarensis* remains (Ethiopia), the recognition of *Australopithecus anamensis* (Kenya), the exposure of very old tool sets (Kenya, Ethiopia), the location and exploitation of new hominoid sites and breccia deposits associated with australopithecines and *Homo* (South Africa).

Clearly my division of time has been somewhat artificial and the decades described do not succeed one another so neatly. And yet, the prospecting trips of the early days with a few locally recruited workmen in no way resemble the expeditions of sometimes 100 people, with four-by-four and heavy-duty vehicles, scrapers, aeroplanes and helicopters which the crazy years saw deployed over two decades; and the portable telephones and computers, GPS (positioning instruments) of the heritage years no longer look anything like the perforated cards and even the most sophisticated compasses of the immediately preceding years.

Chapter 5

LUCY THE FOSSIL
The Story of the Heroine of the Story of the Story of Man

Lucy!

What a surprising planetary destiny for a few hundred little bits of grey bone, 52 of which found their place in a skeleton of a single prehuman very probably of the female sex, a bit more than 3 000 000 years old and a little less than 20 years of age, who was catalogued as "Afar locality 288" and whom we called, first with familiarity, then with affection, Lucy, like the one in the sky.

"We" were a group of about 30 people, researchers and technicians, French, Americans and Ethiopians, united under the authority of a triumvirate – in alphabetical order (suits me), Yves Coppens, Donald Johanson and Maurice Taieb – in an operation which was scientific, geological, palaeontological and prehistorical, the International Afar Research Expedition, between 1972 and 1977.

"Lucy, like the one in the sky", was the Beatles' *Lucy in the Sky with Diamonds,* of which we had the cassette and which imposed itself upon us in the camp to name the little skeleton, when its sacrum, pelvis and gracile nature all at once strongly suggested femininity. It is amusing to recall that the Beatles' title had reportedly been given by the little girl of one of the members of the pop group to a drawing of hers when, home from school, she had been asked by her parents what she had meant to convey in the drawing; but all the same the "coincidence" of the title with the initials LSD may not have been entirely innocent. It is also amusing to say here that many people are convinced that it was "our" Lucy who had inspired the Beatles; some even went so far as to declare to me that it was certainly the fame of Lucy of the Afars that had guaranteed theirs!

Giving names and nicknames to hominid fossils, as one does to the skeletons or skinned corpses in anatomy class, is common practice; it is perhaps a way of distancing oneself somewhat from their reality and from death, and also of getting nearer to them by playing down their importance.

Indeed, the world of human palaeontology knows Mrs Ples, Dear Boy, Cinderella, George, Charlie, John Paul, Abel, Roger and so many others, not to mention Lucy – so that it is never necessary, in professional gatherings, to translate them into inventory numbers or into the binominal Latin nomenclature.

So why then did Lucy, in such a short time, become a veritable symbol of the origins of Man, the mascot of the discipline, the extraordinary messenger of the evolutionist conception of our history, but also an astonishing dream catcher overflowing with lyricism and poetry?

I believe that the fact that we have 52 fragments, hence a certain silhouette, instead of holding, as is often the case, only a skull (Mrs Ples), a face (John Paul), a mandible (Abel), a broken tibia (Roger), or even a tooth ... must have played a part in the reputation of the character whom we could, consequently, better imagine, with her height, weight, carriage and gestures – standing erect, like us, small, slim, quaint in her tree-climbing skills and touching in her awkward gait.

The fact that this fragmentary skeleton was old (3 180 000 years at the last count), ancestral (even if Man is not directly descended from its kind), female, that it had reached its twentieth year (which, for an australopithecine, was very old!) and that it had been given a name (familiar to a large part of the planet) also helped to construct a likeable image; in the collective imagination appeared a young woman – indeed a young girl – paradoxically the grandmother of the whole of humankind and, all at once, so close to us in many features we all share and so exotic in many others. A true legendary character that elicited attraction and curiosity, sometimes even uneasiness.

Like all emblematic personalities, her name was very soon seized to inspire and sometimes name essays, novels, poems, songs and drawings, sculptures and a great deal more.

It would be interesting to count the number of scientific articles the world over which have had Lucy or parts of her as their subject or point of reference for a quarter of a century since we enabled this individual to see the light of day again – for example, at the Collège de France in 1990, I gave a series of nine seminars titled "Lucy in pieces". As we have already seen, many australopithecine remains have been brought to light since the first (found in the quarry at Taung exactly 50 years before Lucy, November 1924 – November 1974!), in other words palaeoanthropologists have known for many years what the skulls of australopithecines, the casts of their brain cases, their teeth, their vertebrae, their waists, their limbs and even the extremities of their limbs looked like. But Lucy's great originality is to have enabled palaeoanthropologists for the first time to observe all these bones or at any rate many of them assembled on a single skeleton, in their respective dimensions, their proportions, their articulations and their relationship.

Lucy is not and has never been, as the media have too often trumpeted, "the oldest woman on earth", but rather "the least incomplete skeleton of one of the oldest pre-humans". But understandably my definition would not have made as pretty a headline as theirs.

I shall now first propose to the reader an anatomical examination of the character (Fig. 6) from head to toe.

She is a small hominid (1 metre to 1,2 metres tall), of modest weight (20 to 25 kg) – a height and a weight that

Fig. 6 – The skeleton of Lucy, postage stamp of the Republic of Ethiopia; Lucy had also been given Ethiopian names, Birkinesh or Dinkinesh.

you, at three years old, will very soon reach, little Quentin – who, in her proportions, displays an upper limb slightly longer than ours in relation to the lower limb, but since this lower limb is short, these proportions must be corrected when compared with ours. One could say, in a word, but therefore with inevitable inaccuracy, that Lucy seems to have long arms because she has short legs.

Lucy's skull is very broken, but other skulls or fragments of skulls from the same Hadar site and other localities nearby in space and time show that it was small, projecting, with marked reliefs and constrictions; it therefore covered a small brain, whose volume has been estimated at less than 400 cc. Casts of its interior were of course made to try to read the convolutions, the sutures, the sulci, the areas and the middle brain vascular network; the occipital, which is pushed towards the back, and the consequent development of the frontal and parietal areas suggest (according to Ralph Holloway, whom I am happy to follow in this) that a structural modification has already set in despite the absence of significant volume growth.

It is in fact very probable that a change in stance and locomotion, way of life and behaviour, resulting from a changed milieu, cannot go without a change within the central nervous system; but since the qualitative and quantitative changes were not simultaneous, the first probably preceding the second by millions of years, it is no simple matter to prove the one, affecting only the content, while the other, affecting both container and content, does not yet appear.

The morphology of the base of the skull, which had also been examined on other subjects than Lucy, does not show the curvature that accompanies the descent of the larynx and points to articulate language. Lucy and her congeners must very certainly have communicated in a very elaborate fashion through sounds, cries, intonations, modulations, signs, gestures, miming, and so forth; you need only have lived your first 18 months, do you not, Quentin, to know how many things it is possible to "say" before you know how to speak – especially when those things are "said" to those who say and understand them like yourself. That being said, it seems clear to Jeffrey Laitman that Lucy did not speak as we speak – with a descended larynx, a sound box placed between this larynx which does not fossilize, a soft palate and a shrunken symphysis, but also with well-developed cerebral areas associated with language.

Lucy's teeth have thick enamel; their inner surfaces are essentially marked by the stigmata of a vegetarian diet. The incisors, relatively developed, especially the first ones, have crowns with tops crenellated by chips and grooves, probably owing to the consumption of branch ends (these being placed in the mouth and the stalks then being pulled out through clenched teeth to strip them of their leaves); the canines are not large; the premolars and molars, on the other hand, are, with a tendency to larger cusps so as to increase their surfaces; even the first lower premolar, in spite of its single cusp (so characteristic of the great apes according to old criteria), shows, in its appearance, its "desire" to molarize – Henri Albertini saw it as having two cusps, so pronounced was its tendency towards expansion. All these side teeth, strongly clenched together, with their grinding character more and more marked from the first molar to the last, furthermore display shiny and very worn facets owing to their consumption of roots, tubers and, inevitably accompanying the latter, earth. The differential wear from the front to the back also shows the lengthy duration of the cutting of teeth, reflecting the increase in the overall growth period, which is itself par excellence the time of learning.

This dental equipment, powerful for the jaw carrying it, is naturally operated by a musculature which is itself the reflection of its function and entails the existence or the persistence of the facial and cranial features briefly referred to above.

The curves of the spine confirm what we learned from the structures of the skull and the position of its foramen magnum; Lucy stood erect. Her pelvis is also particularly significant in this regard: short in its height, it is broad, even very broad (iliac blades, ileum, sacrum and hence the pelvic cavity); the sciatic notch on the other hand is not at all prominent and the hip joints are small.

The femur is short; its neck is long and slender; it is oriented obliquely in relation to the pelvis but perpendicularly to the line running through the middle part of the femur, the femur itself, consequently, being oblique in relation to the spinal axis; at the lower end of the femur, the groove between the ridges is broad and deep; the tibia is short and its ridges are very close together; the foot is short, broad and flat, carrying the weight of the body mainly on its outer edge, and with the first toe diverging.

What is the meaning of all these rather technical data? A short pelvis, that is to say a compacted pelvis, is one that has, among its functions, that of literally carrying the whole part of the body placed above it; this pelvis therefore confirms the erect stance of this australopithecine. The obliqueness of the femur, which tends to bring in the knees towards the body's centre of gravity, also supports these conclusions; the shortness of the pelvis (in height) as well as its breadth also reveal bipedal locomotion, but, shall we say, bipedal locomotion of a certain kind.

On the other hand, the modest insertion of the gluteus maximus on this pelvis suggests that the musculature of the buttocks of this australopithecine was rather like that of the great apes; Lucy did not have a round backside! The widening of the pelvis, the orientation of the iliac blades, the small hip joints and the length and orientation of the neck of the femur furthermore reveal a more efficient mechanism with a large ischio-femoral muscle (external rotator of the thigh and extensor of the hip) and a powerful gluteus minimus (flexor of the hip and internal rotator of the thigh).

The extraordinary footprints imprinted on 3 500 000-year-old tuff at Laetoli in the north of Tanzania, some 2 000 kilometres from Hadar, strikingly confirm, if need be, the bipedalism of the hominids of that period – but, by definition, we shall never know if these footprints were left there by the same hominids as those which Lucy belonged to – and at the same time they emphasize, by their alignment, or indeed their criss-crossing trail, the peculiarity of that bipedalism.

All these anatomical data, whether direct or indirect, and the well-argued reconstitutions that they have made possible, lead us to imagine that the stationary balance of these prehumans was generally not as good as Man's; the instability of the hip, knee and ankle joints could not have enabled Lucy to endure the upright position of any sort, and a fortiori the immobile upright position, for a long time; no question of standing to attention for hours, as the soldier that I was (for 18 months) sometimes had to, on the national holidays of 11 November or 14 July. As for walking, though assisted by powerful extensors and flexors (according to the traces of their insertions on the bones), it could also have been fairly unstable, easily unbalanced and consequently something of a trot, so as to quickly regain balance, with the lower limbs extended and the upper limbs swinging to facilitate the rotational movements of the pelvis and the shoulders around the vertebral axis. In other words, Lucy swung her hips very noticeably!

The dimensions of Lucy's pelvis make it possible moreover to estimate those of the skull of her foetus at full term (its length would have been in the order of 90 mm maximum); these dimensions linked to the anatomy of the pelvis (flattened pelvic strait, small distance between the sacro-iliac and hip joints) suggest a complex obstetrical process – a movement of flexion, deflexion and torsion of the foetus – before parturition in front of the ischium (lower part of the pelvic bones), as in female *Homo sapiens*.

It is interesting to note that it was the constraints of bipedal locomotion which brought about this curious shift from the retro-ischial parturition (behind the ischium) of the apes, well established by a tradition of millions of years, to this complicated, difficult ante-ischiatic parturition, which is also our own. The needs of upright locomotion and those of parturition end up in this case being contradictory to some extent.

The anatomy of the femur, of its joints at the hip (at the top) and the knee (at the bottom) and the angle of its neck has already been mentioned. However, let us return for a moment to the knee, for it represents, apart from the title of this essay, one of the turning points of this body towards a new locomotion. The knee is in effect, in simple terms, the articulation of the femur with the tibia: the groove between the ridges at the bottom of the femur fits over the two ridges at the top of the tibia, the two ridges of the femur riding on the matching hollows on the upper surface of the tibia. Well, this fit is very neat and well adjusted in Man, who displays a narrow groove on the femur and well-spaced ridges on the tibia, while it is much looser in Lucy, who has, as we have seen, a wide groove and ridges so close together that it sometimes appears that there is only one. Man's knee is the stable guided joint of a solely bipedal creature; Lucy's is the unstable joint of an arboreal creature, with a large rotational amplitude entailing limited flexion-extension movements and a shorter stride.

The nature of the ankle, the first joint of the foot, the length, flatness and curve of the phalanges conform to this reading of the bones, their morphology and their articulation; locomotion is clearly bipedal, but unstable, with the feet turned in to carry the weight of the body on their outer edge, and coupled to an ongoing ability to climb.

The footprints of Lucy's southern cousins confirm this anatomy: there is a narrow heel, the foot broadens towards the middle (development of the long abductor muscle of the great toe, as in tree-dwellers), the external edge of the foot supports the body, the great toe is separated from the others (divergence), the little rounded hollows in the footprints indicate toes folded downward. Let us recall for the sake of comparison that a purely bipedal Man's footprint shows a wide heel, the great toe together with (parallel to) the others, all of them short but straight, and a pronounced plantar arch.

The upper limb corroborates the diagnosis of continued arboreal locomotion by displaying first the very pronounced solidity of its three joints, shoulder, elbow and wrist. Then, the prominence of the insertion of the flexor muscles of the scapula suggests the use of these muscles in suspension (from branches, etc.) The humerus, notably with its deep, narrow biceps insertion and its double distal trochlea, points clearly to frequent suspension and resulting stability, a good sound stability without being as inflexible as in the chimpanzee (movements of flexion generally – adduction, abduction and rotation of the arm).

As for the features of the ulna and radius, they complete our mechanical reconstruction of the movements of the upper limb, underlining the importance of pronation and supination, and confirm the general tendency of the system to favour the safety of moving in trees. The wrist does not contradict this search for stability; it articulates a hand with long curved phalanges, comparable, in their grasping functions, with those of the foot.

That, then, is who Lucy was, bodily. The importance of the remains which illustrate her clearly justify her scientific success, but that success grows with the results of the studies revolving around her, which reveal this curious character endowed with double locomotion, yet neither as totally devoted to one form of movement nor as perfectly at ease with the other as may be creatures exclusively practising one or the other.

One of the criticisms which could not but have sprung up was to tell us that we had mixed the remains of a biped (a hominid, at least, it was conceded) and those of an arboreal primate. Fortunately for Lucy and for us, her femur represented the golden hinge between the biped and the arboreal, because that bone, by its obliquity and its articulation with the compacted pelvis, guaranteed that it belonged together with that spine and that skull and, by its unstable articulation with the tibia, ensured its unambiguous attachment to the rest of the lower limb and, by implication, its complementarity to the entirety of that upper limb.

This meant many things: the idea, first of all, that there had not been once and for all, one fine day, a single and unique bipedalism of which we were the heirs, but that that type of locomotion had possibly been born several times, in a parallel, convergent or independent fashion, and that each of these bipedalisms had lived its own story, an evolution;

this also meant that thanks to such beings – the carriers of features from before, still present, and features from afterwards, already present – evolution per se, and the evolution of Man from the primates in particular, received a brilliant demonstration; but it also meant of course that these prehumans, of whom there had been millions for millions of years, never had the status of go-betweens-in-waiting which tends too often to be attributed to them, somewhat condescendingly. Who would think of considering *Homo sapiens* as the intermediary stage in a precarious evolution that he does or does not await?

Lucy showed moreover how bipedalism had come to the hominids by first taking over the essential edifice of the body – skull, trunk, pelvis – before conquering the limbs; from head to toe, and not, as had previously been thought, with apparent logic, from toe to head.

Lucy showed finally that the evolution of the primates towards Man, such as it could be described a postiori, had treated itself to the luxury of a major step forward by setting bipedalism in motion, putting in place the hominid brain structure (and its consequences in the probable perception of many things, thought, communication, behaviour), introducing ante-ischial parturition, the possible preparation of the tool ... In future research, many features that had been exclusively reserved for Man will have to be considered as features whose story must be traced throughout the complete story of all the hominids, and even of the hominoids. Man belongs to natural history.

We, the fathers (Yves Coppens, Donald Johanson and Maurice Taieb, co-directors of the expedition which discovered Lucy in 1974) and godfathers (Donald Johanson, Tim White and Yves Coppens, co-signatories of the 1978 article which baptised Lucy *Australopithecus afarensis*) believed, to some extent, in Lucy's role as ancestor of the genus *Homo*; a certain number of authors still believe in it. I have long since left those "militants", because many of Lucy's features appeared to me to be irretrievably autapomorphic (I have often called Lucy, with deliberate exaggeration, *PreAustralopithecus afarensis* to distinguish her phylum) and because it seemed to me difficult to pass almost instantaneously from such a well-established bi-locomotive structure to that of genus *Homo*'s exclusive bipedalism, the one being practically contemporaneous with the other.

Using the very eloquent data of functional anatomy, those of the climatic, floral and faunal environments (the predators, for example, but also Lucy's competitors), those of the sociology of present-day primates and preferably those among them who are phylogenetically close, it has been reasonably possible – and enjoyable – to describe the socio-ecology of Lucy and her kind. Pascal Picq thus sees *Australopithecus afarensis* living in groups (because their very varied diet allows a better exploitation of their milieu) and observing among themselves tolerant social relations (again, it is the diversity of the food which brings about reduced competition between individuals generally and males in particular and also entails the promiscuity that ensues); he sees Lucy leaving her community around the age of 11–12 years to join another and find a partner with whom she will maintain, within this new group, privileged relations (which may be reciprocated, but probably only to a small extent).

As I have said earlier, the discovery of *Australopithecus anamensis*, with its unstable upper limb and its solid lower limb, is perhaps the right answer to this search for the australopithecine who was the true genitor of the genus *Homo* ... But, in all she has come to symbolize, Lucy has incontestably remained that ancestor of humanity that we believed – then wished, and today still wish – we could see in her when she dresses up in her grandmother outfit: Lucy the tropical founder of humanity, Lucy the African – coloured, matriarchal ... After all, it is not the worst image one could offer humankind of its origin.

CHAPTER 6

LUCY THE SYMBOL
The Story of the Story of the Heroine of the Story of the Story of Man

> I must add that you have found the soul which floats above those millions of generations and which you have called Lucy, that nascent little flame mutating among the australopithecines, that symbol of all the prehistoric ages, which you have so endeared to us.
>
> Marcel Pfister, letter to the author dated 25 April 1989

CHAPTER 6

LUCY THE SYMBOL
The Story of the Discovery and Meaning
of the Story of the Story of Man

I should like to end this book with a hymn to the glory of Lucy, who, in spite of her status as a not very presentable australopithecine according to our criteria of beauty ("Australopithéquette", as a correspondent rather impertinently wrote me one day), has never yet been taken as synonymous with ugly, primitive, uncouth, retarded, backward, as have been and continue to be the ancestors who were to follow her, pithecanthropine, Neanderthal and even Cro-Magnon.

I have just read, for example, these declarations by Daniel Domergue, the proprietor of the Clos de Genteilles, in Siran: "I wanted to be done with the legend of solid, hairy wines, with the 'Pithecanthropine of the Languedoc' image that has clung to us for far too long. Clearly the civilization of the troubadours or of the delicate, quasi-Tuscan architecture of Toulouse had far greater appreciation for cinsault or other light, vivacious cultivars. But how were we to find the missing link?"

Calling someone a Neanderthal is probably still worse, as the pithecanthropine has the advantage of conjuring up a more blurred image of someone very ancient and very distant. Neanderthal, on the other hand, lives here; we have seen portraits, sculptures of him, and even thought we saw him on the screen. There can be no doubt that he is synonymous with stupid, defective, especially since the treatment he underwent in a century and a half of reluctant association with him.

As for Cro-Magnon, the painter of Vallon-Pont-d'Arc and Lascaux, I should have thought he would be spared. But, alas, not at all; somebody, admittedly an American, commenting on the film, also American, *Forrest Gump*, which portrays a simple-minded character, remarkably played moreover by Tom Hanks, described the hero as a sort of "Cro-Magnon"! I heard it myself ... I was outraged!

But let us return to gentle Lucy, beautiful and exemplary, infinite source of inspiration, social phenomenon in a society desperately in need of and hungry for myths, and let us enjoy ourselves wandering through her immense crowd of fans, groupies and true admirers.

To open this catalogue, I should like to start with this elegant and funny poem, which sees Lucy as the mistress of Adam and the mother of Eve. The poem is by a school principal, Marcel Pfister:

Prehistoric Fantasy

It all started with a fatal mistake:
in a transaction, there were two chromosomes
that Adam forgot to write into his genome,
which the Simians found degrading.

Especially when they noticed his deformed head,
the monstrous, deviant, hydrocephalic skull,
the colourless face, the hilarious expression,
the aggressive chin and enormous nose.

And the Council of Primates, in the heart of Africa,
having condemned this incongruity,
wisely and unanimously decreed
that such a biological monster should be banished.

And Adam renounced his Simian ancestor.
He went to inhabit rock shelters,
settled in, invited the spirits,
and decorated his walls with sumptuous frescoes.

He ravished Lucy from her savage tribe,
that australopithecine so skilled in mutation,
that gracile girl, her eyes filled with emotion,
and took her home to keep his house.

Some moons later Lucy gave him Eve.
"Behold the bone of my bones, the flesh of my flesh!"
Adam proclaimed. That child was dear to him,
for she was his fantastic dream come true.

And so he programmed the family Sapiens;
Eve blossomed like a beautiful nenuphar.
Lucy went away to die in the Afar,
keeping her appointment with Yves Coppens.

Adam took Eve on a long walk
towards the Promised Land, inspired by intuition,
aware that he bore within him an immense mission.
So it was that the earthly paradise was opened to them.

The second introduction that I have chosen is the illustrated (Fig. 7) essay, so fresh and full of imagination, written by the second-grade pupils of the Jules-Ferry primary school in Rheims, under the guidance of Agnès Leygnat, the headmistress, and Michel Royer, the educational advisor, dated 1988 and signed by the 25 pupils. In a letter dated 3 March 1989, these children wrote to me: "We have done a lot of work based on your research on 'Lucie' and prehistory in general, and this work culminated in the creation of an illustrated book titled *Lucie*, whose main characters are Yves (a famous archaeologist), in your honour, and of course Lucie" (the parenthesis is in their letter ...).

Fig. 7 – Lucy as seen by the children of Rheims (École Mixte Jules-Ferry – 1988); reconstitution: Jacky Piermay in consultation with Marie-Christine Anciant.

Here is the story:

> Caressed by a ray of sunshine, Yves opens an eye, very surprised not to find himself in his bed ...
> He looks around and wonders where on earth he can be. Yves is a famous anthropologist, director of the Musée de l'Homme, and this unusual decor intrigues him at first. He gets up and surveys the surrounding countryside. In the distance, he can make out volcanoes. All around him there is only tall dry grass with some bare, twisted trees scattered about. A few steps away from him, an antelope is nibbling on a thorn bush while a leopard, on the

branch of a tree, lies in wait. Farther off, he sees a rhinoceros stretched out in the sun. Near to him, a gazelle is grazing. Yves takes a few steps forward to explore the surroundings and finds himself face to face with a dinotherium. Very surprised, he wonders whether he is not dreaming ... he pinches himself and understands ... that he has gone back into the past when he sees that animal which disappeared at least 12 million years ago. The landscape and the animals before him tell him that he is in the savannah, probably in Africa.

Suddenly, Yves is aware that the sun has become very hot and that he is very thirsty. He decides to go and explore the surroundings, hoping to find a waterhole. He heads towards a sort of hill, which is nothing but an extinct volcano ...

From this height, he can distinguish a green spot, which stands out from the yellow grass. He decides to go there. He walks for a long time ... and at last sees water shimmering in the sun. In front of him, some water-birds fly up ... He approaches, happy to reach his goal for he is more and more thirsty and feels his lips getting drier and drier.

All of a sudden, before him spreads a more luxuriant, lively vegetation, and, before his startled eyes, he sees zebras peacefully drinking. Gently ... not to frighten them ... he approaches ...

Standing now on the riverbank, he bends down, cools off his face and then, cupping his hands, drinks greedily.

Suddenly, in the water, he sees somebody's reflection ... Very surprised, he looks carefully, but he cannot quite distinguish, for the water is moving. He waits for a few seconds for the water to be still and then, to his amazement, he clearly sees ... a creature behind him. The being watching him is neither animal nor human ... Yves quickly turns round to get a better look at this character, but the frightened creature flees as fast as its legs can carry it. Yves stands dead still, he can see nothing now ... the creature has vanished into the tall grass ... Intrigued, he decides to look for what seemed to him to be a child, but of what race ...? So he sets out after the creature, following the traces it has left behind it, for the grass has been trampled where it ran. Suddenly, nothing ... He stops ... looks in all directions ... sees only a bush. He advances noiselessly, goes around it ... in vain! He decides to part the branches and that is when he sees the creature huddled inside.

Seeing him, it gets up and Yves discovers with astonishment that he is face to face with an *Australopithecus afarensis* ... At that moment, he understands that he has gone back 2,5 million years in time.

The character before him is a young girl, not very tall, about one metre twenty, her face is flat, her forehead small, her eyebrows thick and bushy, her cheekbones prominent, her nose flat with large nostrils, her jaw projecting forward, she has a large mouth with thick lips, her shoulders are round and low, her limbs muscular, her arms seem very long and her pelvis, on the other hand, is very narrow.

Then Lucie – it is the name he immediately gives her – signals to him. He does not understand. She comes nearer, takes his hand and leads him through the savannah. They walk for a while in silence and arrive behind a semi-circular stone wall. She pushes him gently to make him sit down and then makes a sign; he thinks he understands that she is intimating that he wait. She disappears for a few minutes and he sees her coming back laden with roots and berries. She in turn crouches down, holds out to him a handful of berries and some roots and starts nibbling some of them. He is so hungry that this modest meal seems like a real feast to him. When she has had enough, Lucie plucks up courage, she draws nearer and nearer, timidly touches his clothes, his shoes, runs a wary finger across his face, notices with wonder his watch, listens to the tick-tock. Yves takes out his lighter, lights it and Lucie immediately tries to catch the flame. She burns her fingers and runs off shrieking into the savannah. Yves chases after her for he wishes to tell her that it is not dangerous and to show her that it may be used provided one takes care. Lost in thought, he does not see a branch on the ground, trips, falls heavily, bumps his head on a stone and faints ...

At dawn, he regains consciousness. His head hurts. He raises his hand to his forehead and realizes that he has a cut that is bleeding. As a result of the impact, his vision is slightly blurred and he wonders whether the men surrounding him are real or the fruit of his imagination. He rubs his eyes but it is not a dream ... In the middle of these men with their threatening expressions, wielding hand axes and clubs made from antelope bones, he sees Lucie advancing slowly towards him and screaming. She shows her burnt fingers and Yves understands that she is asking him for his lighter. He takes it from his pocket and lights it ...

The group of men, stupefied, draws back. Yves, so as not to frighten them more, blows out the flame and shows them his lighter from far. A man breaks away from the group, comes towards him cautiously, touches his face and hands and holds out his fingers towards the lighter. Yves gives it to him ... With apparent cries of joy, the one whom Yves supposes to be the leader waves the lighter around then, turning to him, places a hand axe in his hand and folds his fingers closed over it. The eyes of the australopithecine then rest on the ring of the explorer and he tugs at it. The whole group draws nearer and all of them give Yves stone tools, animal teeth and bones. They surround him and start a sort of dance. Lucie takes his hand and draws him into a wild dance. The researcher realizes that the group has just adopted him ...

Suddenly, a man approaches the group, making sweeping gestures, incomprehensible to Yves but apparently very clear to the others. They rush into the savannah pulling Yves with them. They arrive in an arid space and see two men fighting with hyenas to wrest from them a piece of meat torn from the carcass of an antelope. Farther on, women are calmly gathering roots and berries. The man who appears to be the chief sits down and signals to Yves to do the same. Straightaway, men and women come running up to them, bringing food, which they serve to them on animal hides. Not to anger them, Yves forces himself to eat a bit of everything. He decides to light a fire to show them that it is possible to cook the meat but, when the chief hands the lighter to him, he sees that it has only a small trembling flame left which soon goes out. He cannot manage to relight it. Seeing that the lighter no longer works, the men become angry, throwing stones at him and Yves runs off, taking cover in the long grass. After a while, exhausted, he stops, lies down and, dead tired, falls asleep. He is still holding the hand axe, the chief's gift ...

Suddenly, someone touches him on the shoulder ... Convinced that it is Lucie coming to fetch him, he opens an eye and discovers, amazed, a woman in a grey coat ... It is the cleaning lady of the museum who says to him: "Sir, it is eight thirty, you must have fallen asleep. In half an hour the museum opens ..." Sheepishly, Yves gets up. He has dreamed that beautiful adventure ... He fell asleep while studying Lucie's skeleton and the objects found near to her. In his hand he is still holding the hand axe, the last evidence of his dream ...

Apart from the rather premature extinction of the dinotherium, and apart from the fact that Lucie and her people are here equipped with hand axes and have dwellings built of stone, which they could not have had, and that the Musée de l'Homme opens at ten and not nine, there is nothing to change in this sweet dream, which I should have loved – and should still love – to have.

In the theatre world, it was Michel Arbatz who brought our character to the stage, in *Ma nuit chez Lucy* (My Night with Lucy). "Through the medium of a musical and theatrical fiction," he writes, "I have assembled the words of scientists, sounds and songs which, each in their own way, question the creation." A refrain sung by Yonval Micenmacher, Jean-Luc Michel and Véronique Feller (in the role of Lucy) at the opening performance of the play in 1994 in Avignon and Montpellier, said:

Oh, Lucy
Where are you hiding, say?
In what hole in time
At the bottom of our dreams
Naked as the wind
You see we are waiting for you, here
You are my double, my trouble too Lucy
We are still alike, we stammer, Lucy
Two fireflies in the burning air flying, Lucy …

In the bipedal rap, the three actors named above and two others, Patrick Leroux and Michel Arbatz, shouted in chorus:

If you want to go straight ahead, go ahead,
Lift your head, athlete,
Whether you're a pithecanthropine or an australopithecine
Straighten up your spine with its double "S"
…
Walk, walk, walk, walk, walk, walk, walk, walk, walk Man!!

The brilliant theatre personality, Alain Germain, always full of ideas and humour, before publishing *Les Origines de l'Homme ou Les Aventures du professeur Coppensius* (The Origins of Man or The Adventures of Professor Coppensius) in 1997, had staged it as a show in 1991, presented at La Halle Saint-Pierre during an exhibition on the same subject. Two student volunteers are sent back in time as an experiment by an eccentric professor and his assistant Rouletonos. "I think we can now lift the veil which separates us from your great-great-great-grandmother," says Professor Coppensius to his assistant when the students' voyage reaches the 3-million-year mark. "Notice how the body has evolved ... But the miracle is in the equilibrium of this body which stands erect on two legs ..."

In the cinema world, Daniel Vigne courageously embarked on a story, in the burlesque mode, inspired by our East African research and the discovery of Lucy; the film was called *Une femme ou deux* (A Woman or Two); Gérard Depardieu was the palaeontologist, somewhat excessive in his enthusiasm perhaps, but endearingly absent-minded. As for Lucy's double, the skeleton called Laura, Depardieu discovers her somewhere in the Massif Central, in the heart of France.

Still in cinema, a more recent film (1998) by Janusz Mrozowsky, *La Revanche de Lucy* (Lucy's Revenge), calls upon the most ancient African ancestors to come to the aid of populations subjected to the whims of a dictator.

Literature, in both prose and verse, has also been touched by the heroine of our discipline.

In March 1995 Henri Lahaye published this sweet cry of hope:

Lucie

This morning, hearing an astronomer
Calmly speaking of billions of suns
I gave up on shaving last night's beard.
Even when clean, how small man is, a mere trifle ...

His inner pain, of love, of life, of being born,
His monthly paycheck, his end-of-year parties,
His lows, his throes, his debates, his jeers,
All that is just a lot of rubbish ...

I was knotting my little destiny at the end of a rope
When the scientist spoke of you, Lucie,
Our ancestor of more than three million years.
Far-away grandmother, you saved my life!

You restored to me a little of my importance.
Suddenly I saw thousands of women giving birth
So that the world could finally attend my birth!
Epsilon-man, thanks to woman, you are no mere trifle.

In *Mémorable planète*, Luc Estang also has a good time:

> ... Long-armed anthropoids
> fat gorillas and chimpanzees
> auburn-haired orang-utans,
>
> Then a break-up between all these
> and the ancestors of Lucy
> who bears us the evidence.
>
> Three to four million years
> Ethiopian of the beautiful teeth
> She lived, already standing on her own two feet
>
> From Southern Africa to Cro-Magnon ...

Pierre Pelot, in complicity with me, attempted in the very elegant *Rêve de Lucy* (Lucy's Dream), written in 1990, to slip himself into the character.

> She was the third and last adult female of the group
> – before, when she was little (these are the things she remembers), she was not as strong but could run

faster and jump higher, like those others now who came after her. The third and last female, that means that, first, in front of her, there is the female who follows the male walking ahead and who walks fastest, and the other female, the one who sees only once. She has always seen them, and they are still there. There were other females before, but they stopped walking, at one time, at another time.

She would often remember, but not all the time. Of course, there were more propitious moments. Sometimes, it was what she saw, whatever lay before her eyes, that conjured up other buried pictures. The phenomenon was strange and always surprised her, causing a mixed sensation of pleasure and fear – and she understood neither pleasure nor fear, she could not explain or find any reason for this reaction. And yet they were images that had crumbled to dust, vanished like when the shadow fades at the end of a gesture while the light of the sky wanes; they were images that had *disappeared*, and yet would surge up inside her again. Only for her? Did all the others know those same images, those very ones, or did they each possess their own?

What once is, after it has passed, does not become carrion and mushy rot, like a tree or any other living thing; no: there are the images, they ask only to be seen, gathered, adopted. Images float in the wind, on the breath of the one who blows hard and is never seen; images take refuge in the stomach of those who eat them; they need only cling to roots, leaves, everywhere. Perhaps you even swallow them when you breathe.

In a short story called *Le choix de Lucy* (Lucy's Choice), written in 1993 for a competition organized at the Atomic Energy Commission, where he works, Jean-Luc Sida leads Lucy to suicide: "I am very sorry," he confided to me on 29 November 1993, "and I sincerely hope that it has not shocked you. Try as I might to turn the problem around, I could not construct the story any other way."

Lucy, through telekinesis, leaves microscopic messages on the facets of a crystal in which she intimates that she is weary of being an australopithecine, exhausted by having to communicate despite herself in this telepathic fashion – a type of communication which natural selection, proving her right, has moreover since discarded, along with the prehumans!

As for the novelist Andrée Chédid, in an important and sumptuously lyrical text, *Lucy, la femme verticale* (Lucy, the Vertical Woman), she drowns her so that Humanity should not happen!

The first part, Lucy's appeal to her descendants, is full of pathos: "It seems that my defunct body, with its embryo of a soul, its film of spirit, inhabits and haunts yours. And it appears to me that the destination of my flesh is your flesh; that my gesticulations spread out to your gestures, that my simpering chatter opens into your laughter, my little hops into your strides." But the author decides to kill her – "I shall stifle the human race and its perverse destiny ... How much chaos we shall have avoided! How many monstrosities we shall have abolished, in exchange for that so brief, so perilous gift that will have been our existence!" – before exclaiming in a song of resignation and despair: "Let the universe roll out, as it must, its history. Let Lucy accomplish her purpose ... By the grace of Lucy, I shall exist, you will exist, we shall exist."

Even the novelist Erik Orsenna rounded up all the guitarists of all the ages, one day in the heart of the Rift Valley, in our Omo camp, for an immense concert bursting with notes, colours and lights, in homage to Lucy:

> The pink helicopter flew over the lake, defied the flamingos, sought in vain some crocodile and finally landed.
> "Awfully sorry we're late," exclaimed George Harrison, the first to alight ...
> Paul and Ringo appeared and, a few seconds later, John Lennon, more Asiatic than ever, in a crew-neck tunic and small round glasses, like a mischievous mandarin ...
> Then, with adolescent insolence, they started to play, taking no notice of the others.
> And so it was that *Lucy in the Sky* rose again in the African sky as night fell ... the twinkling African night, alive with stars, the 3 million candles in Lucy's cake.

And I shall never forget the crazy extravagance of the late Pierre Schaeffer; in *Faber et Sapiens*, this is how he describes – rather irreverently, I must say – the activities of our lovely mascot:

> In the long grass of the savannah fumbles an *Australopithecus africanus*, of the *gracilis* variety, unless it's an *Australopithecus afarensis*, the improbable transition between the australopithecine and the ramapithecine, in short, it's Lucy (Aunt Lucy), schnoozling away in Ethiopia.

And in a note he explains that "to schnoozle" (*bassotter*) is an expression meaning "to attend to irrelevant business"! Then follows a charming description of Lucy's pickings and thoughts, both taking place at the same time, until her return home, where her "guy", "a rather hefty Gracilis", and a brood of squealing little nephews wait for her (or do not).

> "So, were we good then?" asks Aunty Lucy, "otherwise we'll call Cousin Robustus who'll take the little Graciles deep into the forest." These retro remarks are greeted with loud booing.

It cannot be said that there was no atmosphere in Lucy's house and Pierre Schaeffer's head.

As for yours truly, apart from the multiple complicities that I have greatly enjoyed honouring, I was induced to pen a single, very short love story between Toto (whose name is a wink at the Tautavel cave in France) and Lulu (guess who) – a thwarted love story, because here both time and space were involved. It was at the request of Laurent Broomhead for one of those light and pleasant summer holiday programmes.

The two- and three-dimensional representations of Lucy are of course legion. Artists, regardless of the more or less identical information available to them, have each managed to project onto their work their own individual talent but also their own phantasms and those of their particular culture (Fig. 8.1, 8.2, 8.3 and 8.4). It is amusing to see for example at the Commonwealth Institute in London a very thin Lucy with pendulous breasts (Fig. 8.2) and at the Muséum in Geneva a very round Lucy, all the more so since she is pregnant (Fig. 8.1), at the Musée de l'Homme in Paris a hairy and modest Lucien, at the American Museum in New York a Lucy and a Lucien every bit as hairy as the Paris Lucien, he holding her tight, by the opposite shoulder (Fig. 8.4). And, certain liberties having been taken with the scientific data, it is no less amusing to find an Italian Lucy with a tiny waist, the type who could have found herself being harassed in a laboratory, and a French Lucy with large earrings and a leopard-skin drape, whom one might eagerly have wished to assist as an informant in the bush! All this would deserve an album one day.

But this "sort of charming old grandmother" (as Yaguel Didier movingly calls her) also lent her doubly famous name, thanks to the Beatles but also to her own celebrity, to a complex, a cup, a syndrome, an experiment, and now also an effect.

The "Lucy complex", named and described by Claude Lorin, a professor of clinical and pathological psychology at the University of Rheims, is related to schizophrenia. "I was staggered," he wrote to me in a letter dated 19 November 1992, "by your mythical reconstruction of Lucy's (internal and external) world: it seems extraordinarily accurate to me or, in any case, plausible and it corresponds strikingly to what certain schizophrenic or hallucinating patients experience, disorientation in time and space, or, like a three-year-old child, no notion of yesterday and tomorrow. My question is simple: how can one explain such a coincidence, in your opinion? Why is the symbiotic world of the first hominids so close to psychotic delirium and poetry? Or the opposite: why is the world of the insane so close to that of the first humans? I have the impression that my patients are the witnesses of a vanished world." And in a letter of 18 December 1992: "I intend to thank you in my book for enabling me to understand that there exists a phylogenetic transfer which refers us back, at every birth, to the original world of the very first hominids [...]. I believe [...] that, in the first months of life, we all go through the 'Lucy complex'".

The "Lucy syndrome" is due – if I may say so – to Dr Pierre Pilardeau, of the Medical Faculty in Bobigny, Paris, a specialist in sports medicine. In a seminal article, he and four colleagues define this syndrome as a consequence of the shortening of the pelvic muscles, a modification induced by straightening the body to the upright position. In a 1992 medical doctoral thesis, inspired by this original hypothesis, a student of Pierre Pilardeau, Sylvain Dionnet, shows for example that 62% of 170 football players examined suffer from at least one of the pathologies of this syndrome. The football player, who has to burst into a sprint about 100 times per match, each of 3 to 6 seconds' duration, in a semi-flexed position, repeatedly causes the retraction of his pelvic muscles, often straining them. In other words, "the physiopathology of the Lucy syndrome results from the non-adaptation of bipedalism to fast running. To correct this handicap, Man adopts a dynamic attitude which tends to revert to that of his origins." The consequences of the syndrome are numerous: contractions and pain in the back, pulled and strained muscles, ruptures and muscular troubles of the abdominal and pubic areas, stretching and tendinitis of the adductor muscles of the thigh, displacement of the kneecap, and so on.

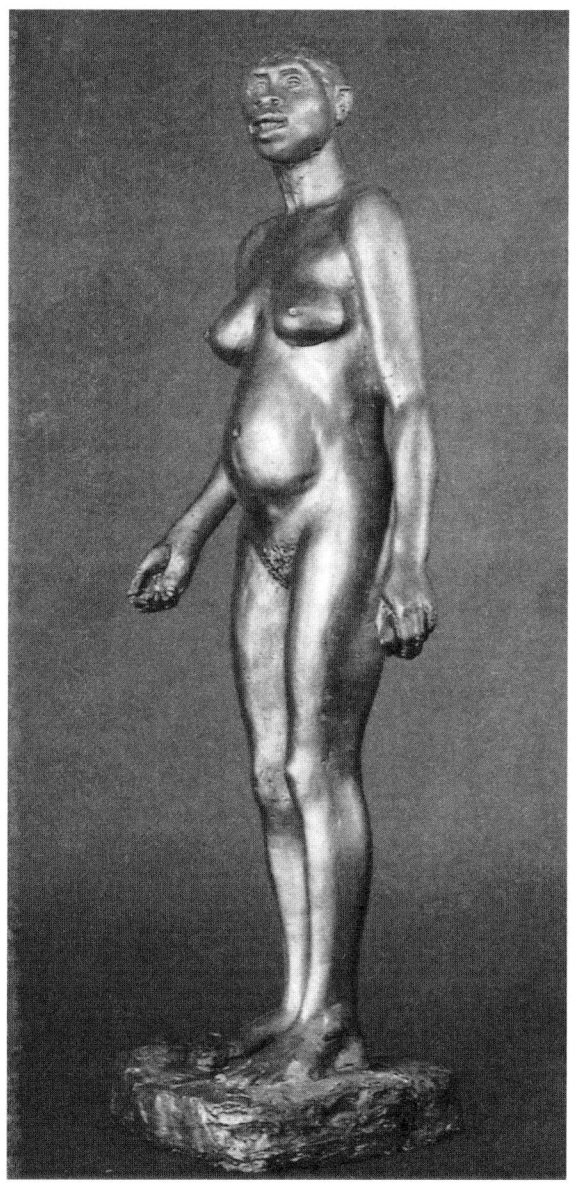

Fig. 8.1 – The Swiss Lucy, Muséum d'Histoire Naturelle, Geneva; reconstitution: Gérard Métral and Olivier Bindschedler.

Fig. 8.2 – The English Lucy, Commonwealth Institute, London; reconstitution: Derek and Patricia Freeborn, Freeborns Studio.

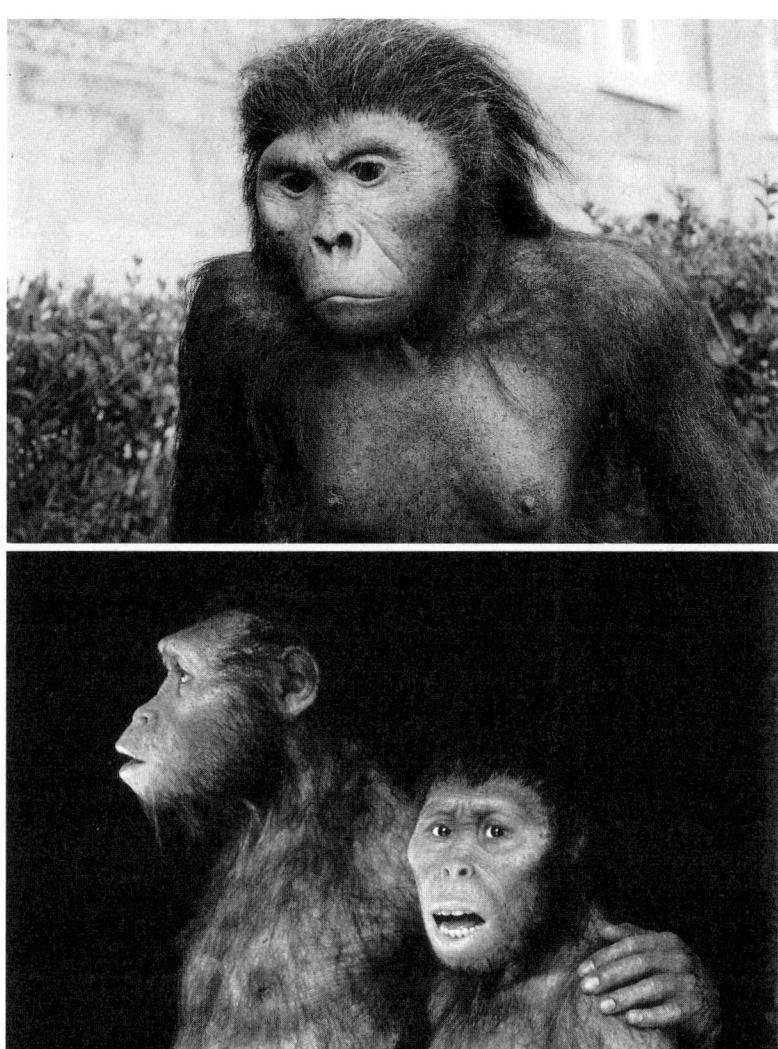

Fig. 8.3 and 8.4 – Two 'American' Lucies and one Lucien.
Top: Muséum National d'Histoire Naturelle, Paris;
reconstitution: William Munns.
Bottom: American Museum of Natural History, New York;
reconstitution: John Holmes.

Donald Johanson told me a few years ago, when I was at his home in Berkeley, that there used to be in Addis Ababa a Lucy Memorial Cup for football, but I do not think the said cup was meant to celebrate the recently identified syndrome that could threaten competitors for the cup.

The experiment called the "Lucie experiment" is the work of four researchers of the Laboratory of Laser Physics of the University of Paris XIII-Villetaneuse, Christian Miniatura, Jacques Baudon, Jacques Robert and Olivier Gorceix, and refers to an elegant manipulation in atomic interferometry; the relevant apparatus "uses a beam of polarized hydrogen atoms that are sent through a region where an inhomogenous magnetic field prevails, altering the spatial movement of the atoms". Good old Lucy!

As to the "Lucy Effect", you are of course in the process of becoming aware of it.

Before I try to alleviate the obsession of the above-named effect, let me not forget to mention the works of scientists, those players of the first or second generation, as they say in computer science (Donald Johanson, Germaine Petter, Brigitte Senut, Pascal Picq), the works of journalists (Yvonne Rebeyrol), the exhibitions (the abbey of Saint-Gérard-de-Brogne), and so on, which had Lucy as their title.

I shall conclude with three authentic scenes, one associating you, little Quentin, with Lucy and me, another probably associating me with her but perhaps with someone else, and the third dissociating me totally from her, which goes to show, if need be, that the future belongs to the past.

It was in the rue de la Roquette; it was raining. You were in your pushchair, Quentin, and I had raised the hood. Only the two of us; we had just been to see the paediatrician, who had performed on your heel – without so much as a peep out of you – one of those vaccinations inflicted upon little people of your age and we were going up towards la Bastille on the pavement, very narrow at that point and on the right-hand side of the street when going in that direction (the opposite direction to the traffic). There were many cars, as there often are in that quarter, noisy, slow as always on rainy days, when there appeared, "interstratified" in the queue, a huge lorry even more arrogant with its screeching brakes and multiple accelerations than the other vehicles, which were hardly discreet themselves. Suddenly a loud hiss immobilized the monster – this time for no apparent reason – in a sort of spasm that shuddered through its whole length, and from the left front window, on our side, a good third of the beaming driver emerged pointing a determined finger at your modest vehicle and shouting in a happy roar of laughter: "Is that Lucy?" As he pulled off again I yelled back above the din: "No! It's her little brother!" I'm afraid, Quentin, that you were taken that day for your father's Cinderella (as Wiktor Stoczkowski says).

The second scene takes place in a taxi. The driver and I were heading towards Roissy at the break of day; I was to take one of those busy-bus aeroplanes to go and tell about the evolution of Man somewhere. I found the driver likeable but absent-minded and restless, and at the same time a little slow for the time I had left and I was, I must admit, a bit on edge. He kept turning round, and would look at me strangely in the rear-view mirror, with a trace of a smile that never left his face. But the suspense suddenly ended; he gave a little bounce of satisfaction on his seat, turned right round – I got a real fright – and, delighted by the victory of his memory struggle, dealt me a: "And how is Julie?" Not knowing whether he had really recognized me and whether he really meant Lucy – after all, Julie could have existed, and he could have been mistaking me for someone else – I took a while to reply, which made him visibly uneasy, then delivered him by venturing: "Oh, you know, still as skeletal as ever!" Well, I must have made the right assumption, and so must he, as he at last seemed satisfied, pleased with himself and pleased – I suppose – to be transporting me. He finished the trip radiating contentment, at the appropriate speed, and in good time. I caught my flight.

The third story, finally, sets me resolutely apart from Lucy's destiny; glory is ephemeral and I should have expected this separation to happen one day or another. Now it is done! I wish Lucy a long career! How can one be jealous of what one has brought back into the world? In short, one fine morning, a telephone call from Jean Lallier, author, director, producer and friend, invited me to collaborate on the production of a television film on the story of Man. A few months later, the 52-minute piece was in the bag, filmed in East Africa and Paris, edited, mixed, ready for delivery; Jean Lallier, very kindly, out of courtesy to her and friendship for me, titled it *Yves, Lucy et les autres* (Yves, Lucy and the others), and the film was programmed on one of our national channels. The television magazines, naturally informed in advance, wanted to give particular importance to that document – at least some of them did – by completing the few lines devoted to it – its title, date and showing time – with an illustration. And that is when a young woman – judging by the voice – from one of those magazines, which will remain nameless, called my secretary to ask for ... a photograph of the two skeletons!

Conclusion

Strange indeed, this story of yours, Quentin, this story of matter which grows in size and wisdom for at least 15 billion years of increasing complexity and organization, 15 billion years which have left proof in each of our bodies of the great stages of this extraordinary journey. Take a good microscope and look at your skin, you will see there the same atoms as in the stars; reduce the magnification a little, you will make out large molecules like those floating in space or clinging together between the sheets of clay beneath the waters 4 billion years ago; a much more modest microscope will give you wonderful pictures of those cells which appeared on Earth 2 million years ago, nicely woven together as there are a billion of them. A simple X-ray of your thorax will show you the internal skeleton which has been living inside you for only half a billion years and, between the vertebrae of its pivot and the ribs of its cage, the lungs that you fortunately developed 400 million years ago; if you run your tongue along your teeth, you will see that they are different (even the 20 milk teeth that you have now), in front and behind, just as they had already come to be in your ancestors, the mammalian reptiles, 300 million years ago, and if you now run your fingers along your skin, you will feel the soft down which, 200 million years ago, elegantly replaced the scales that you wore for such a long time. Take a pencil to write to me and you will see, in three-dimensional colour, your hand opposing your thumb against your fingers to pick it up just as any little ape would have done 50 million years ago if it had had pencils and something to write. But, unlike the ape, you will be very much clumsier picking up the same pencil with your foot between your big toe and the others, which have been parallel in you for the last 4 million years. For a few million years you have also had the astonishing possibility to be conscious of all this, the privilege to be amazed by it and the luxury of reflecting on it – and all this because of the volume of the brain that you now have in your head.

Thus we are unquestionably part of the Universe; but what then is this being called "human", a being which is insignificant on the phyletic tree of life, a speck of dust on its tiny planet, 150 million kilometres from its star, among billions of planets and billions of stars, in a galaxy floating among billions of others; what then is that being next to the infinite spaces and times of this Universe that it sees? And at the same time, what is more complex, better organized, more elaborate than a human being, the sole receptacle of thinking matter, the sole holder of freedom, which knows that it knows, and uses that knowledge to understand the world, understand how it works and shape its working to its benefit? Well, the human being is a vertebrate, a mammal, a primate, humble and obedient but powerful and responsible.

When the question arises as to the purpose of prehistory, the most important answer is this: prehistory puts Man in his place; it helps us to understand who we are, the way we became what we are ("how") and the reason we became what we are ("why").

Confident in this perspective, prehistory is moreover the only discipline capable of expressing itself credibly on the future evolution of humanity; Man is a living being; and as no living being has lived for a long time in a stable moulded "envelope", there is a good chance that Man will not escape this inexorable law and that in a few thousand years he will not look like he does now. But since it is true that the day he started striking one stone with another to change the form of the first, he started to change the world and has never ceased to do so, he has woven a screen of knowledge between the demands of the environment and his own body; in the place of an instinctive biological reaction, little by little, and with some arrogance, a conscious, cultural reaction has thus settled in. Man's biological evolution has consequently slowed down, compacted, halted perhaps. As long as culture in this way comes up with the answers to the aggressions threatening the body which carries it, we are winning time from biological transformation and developing our knowledge and our mastery of the world at a dazzling rate; but if it happens one day that we cannot parry, biological evolution could reclaim its rights, its powers, its duties, for a while or for a very long time, and see to it that we adapt to new worlds and become another species or another genus, obviously much more complex, much better organized.

As we have seen, it has taken many years to understand the progression of the story as told here, that is to say as it is understood today – which clearly does not mean that it will necessarily be understood that way forever more. Many fossils, which cannot lie, have been extracted from the archives of the planet in the course of these 170 pioneering, crazy or heritage years; they are there, powerful witnesses, and remain open to many more readings.

Let us pay homage here to all those palaeontologists, anthropologists, palaeoanthropologists, prehistorians, archaeologists who, fossil after fossil, stone flake after stone flake, have made it possible to draw the lines that come together in this coherent and wonderful story that you have just read.

But very clearly, in the scientific attempts to find and reconstitute the origin of Man, there is also an extraordinary philosophical, poetic and even dramatic dimension: the slow emergence of the hominid from the animal world, for example, the difficult breakthrough of his consciousness, the heavy rising of his body to the erect stance and the touching instability of his first bipedalism, the clumsiness of his first attempts to shape stone and the moving tenacity to improve them. Palaeontological discourse entirely contains within itself this dimension, but expresses it very rarely, very partially and very badly because it does not know how to and sometimes does not wish to. Writers, poets, actors, painters and sculptors are fortunately there to extend the data and their interpretations and open the way to broader speculation, to phantasms, inventions, creations, dreams ...

Let us pay homage to them and let us pay tribute to their imagination and talent which lift us up on their wings away from our beds and our magnifying glasses to offer us another, warmer, more luminous approach to our so skeletal subjects.

Finally, let us pay homage to all those prehumans and anonymous humans who advanced prehumanity and humanity to the level of our own; the australopithecine or the Man who "flaked" the first stone, the one who told his neighbour who flaked the second, the one who reflected and then improved on the next flaking, the Man who thought about the story of Man and constructed the first scenarios while thanking the heavens, the one who copied nature's symmetry, the one who first kindled a flame and mastered fire, the one who picked up the first fossil to start a collection, the one who left the first graffiti in a cave, the one who threw down the first seed to see what would happen ...

But homage, too, to those prehumans and humans who, happily for us, were preserved in the entrails of the sediments, against predation and pressure, erosion and dissolution, right up to our laboratories, and who at the same time were given new names, Abel and Dear Boy, George and Cinderella, Claire and Lucy ...

Well, Lucy, dear Lucy, please receive, at the end of this journey, the last of my homages, a salute to your age, a wink at your beauty and a book to your memory; but the last kiss will, after all, be for you, little Quentin.

Glossary

Abduction: divergence (e.g. of great toe in relation to the other toes; turned out or in).
Adduction: parallelism (e.g. of the great toe to the other toes).
Angiosperms: flowering plant with enclosed seeds (e.g. apple tree, but also wheat, the oak or the cactus).
Apomorphic features: new features, derived features.
Autapomorphic features: idiosyncratic apomorphic features belonging only to a given form or species and shared by no other.

Beringia: formerly dry land linking Alaska and Siberia, now covered by the waters of the Bering Strait.
Biface: bifacial stone tool, i.e. worked on both sides. Most hand axes are bifaces.
Breccia: sedimentary rock, conglomerate of angular stones, etc., cemented by lime.

Cainozoic: geologic era covering the Tertiary and Quaternary (the last 65 million years).
Cretaceous: geologic era extending from 144 to 65 million years ago.

Dinotherium: huge elephant-like animal from the Miocene era.
Diploe: layer of spongy bone inside the bones of the skull.
Distal/Proximal: (referring to part of a bone) situated farthest away / closest to the skull.

Endocast or *endocranial cast:* a cast of the interior of the skull.
Eocene: geologic era extending from 56 to 34 million years ago.

Foramen magnum: hole at the base of the skull through which the nerves of the spinal column ascend into the brain.
Foraminifera: unicellular marine organisms with perforated shells.

Gymnosperm: plant with seeds that are unprotected by seed vessels (e.g. conifers).

Isotope: one of two or more forms of an element differing from one another in atomic weight, and in nuclear but not chemical properties (i.e. with different numbers of neutrons).

Lapilli: stone fragments ejected from volcanoes.

Miocene: geologic era extending from 24 to 5,5 million years ago.

^{16}O, ^{18}O: isotopes of the oxygen atom weighing 16 and 18 atomic mass units.
Oligocene: geologic era extending from 34 to 24 million years ago.

Phylogeny: (history of) evolution of animal or plant type; here: "family tree" of the filiations or *phyla* (see *phylum*).
Phylum: major division of animal or plant kingdom, containing species having same general form (plural: *phyla*).
Pleistocene: geologic period extending from 1 800 000 to 10 000 years ago.
Plesiomorphic features: ancient or "primitive" features; inherited features.
Pliocene: geologic period extending from 5,5 and 1,8 million years ago.

Sulcus: furrow on the cerebral cortex (outer surface of brain) (plural: *sulci*).

Trabeculae: fibres or fascicles of bone tissue that organize the internal structure of the bones.

BIBLIOGRAPHY

As this type of essay often dispenses with the need for the sacrosanct scientific bibliography, the author, only too happy to take advantage of this tolerant tradition, has retained only the references of the least accessible texts (theses or press), some books of fiction and works of popularization.

BACON, Anne-Marie. *Les os longs du membre pelvien chez les primates miocènes et plio-pléistocènes : morphologie fonctionnelle, taxinomie, phylogénie*, University of Paris VI, doctorat d'université, 1992.
BARRIEL, Véronique. *Les relations de parenté au sein des* Hominoidea *et la place de* Pan paniscus *: comparaison et analyse méthodologique des phylogénies morphologique et moléculaire*, University of Paris VI, doctorat d'université, 1994.
BERGE, Christine. *Biométrie du bassin des primates. Application aux primates fossiles de Madagascar et aux anciens hominidés*, University of Paris VII, doctorat de troisième cycle, 1980.
BERGE, Christine. *Effet de taille et adaptation à la locomotion terrestre chez les primates : analyse multidimensionnelle du pelvis (primates catarrhiniens, carnivores). Implications dans l'évolution des hominidés*, University of Paris VII, doctorat d'État, 1988, and *Cahiers de paléoanthropologie du CNRS*, 1993.
BESSIÈRES, Michel. *Figaro Magazine*, 18 July 1998.
BRAGA, José. *Définition de certains caractères discrets crâniens chez* Pongo, Gorilla *et* Pan. *Perspectives taxonomiques et phylogénétiques*, University of Bordeaux I, doctorat d'université, 1995.
BRAUN, Marc. *Applications de la scannographie à R X et de l'imagerie virtuelle en paléontologie humaine*, Muséum National d'Histoire Naturelle, doctorat d'université, 1996.

CHÉDID, Andrée. *Lucy, la femme verticale*, Paris, Flammarion, 1998.
COPPENS, Yves. *Le Singe, l'Afrique et l'Homme*, Paris, Fayard, 1983.
COPPENS, Yves. *Pré-ambules, les premiers pas de l'Homme*, Paris, Odile Jacob, 1988.

CORDY, Jean-Marie (ed.). *De la bactérie à Lucy, Van Bacterie tot Lucy*, Abbey of Saint Gérard-de-Brogne, 1994.

DELOISON, Yvette. *Étude des restes fossiles des pieds des premiers hominidés : Australopithecus et* Homo habilis. *Essai d'interprétation de leur mode de locomotion,* University of Paris V, doctorat d'État, 1993.

DIONNET, Sylvain. *Le Syndrome de Lucy chez le footballer,* University of Paris VII, medical thesis, 1992.

ESTANG, Luc. *Mémorable Planète,* Paris, Gallimard, 1991.

FIZET, Marc. *Biogéochimie isotopique (^{13}C et ^{15}N) du collagène des vertébrés : contribution à l'étude d'un paléoécosystème anthropique du Pléistocène supérieur (Marillac, Charente),* University of Paris VI, doctorat d'université, 1992.

GALICHON, Valérie. *Étude de la structure trabéculaire interne de l'ilium des premiers hominidés d'Afrique du Sud (analyse d'images digitales),* Muséum National d'Histoire Naturelle, doctorat d'université, 1997.

GARCIA, Renée Angelina. *Application de la tomographie informatisée et de l'imagerie virtuelle à l'analyse quantitative de la structure des parois crâniennes,* University of Bordeaux I, DEA, 1995.

GERMAIN, Alain. *Les Origines de l'Homme ou Les Aventures du professeur Coppensius,* Paris, Hachette, 1997.

GILISSEN, Emmanuel. *La cérébralisation chez les singes du Nouveau Monde et spécialement le genre cebus : un modèle pour l'hominisation,* University of Paris VI, doctorat d'université, 1992.

GOMMERY, Dominique. *Le rachis cervical chez les primates actuels et fossiles. Aspects fonctionnels et évolutifs,* University of Paris VII, doctorat d'université, 1995.

GRIMAUD-HERVÉ, Dominique. *L'Évolution de l'encéphale chez l'*Homo erectus *et l'*Homo sapiens, University of Aix-Marseille I, doctorat d'État, 1991, et *Cahiers de paléoanthropologie du CNRS,* 1997.

HUBLIN, Jean-Jacques. *L'émergence des* Homo sapiens *archaïques : Afrique du Nord-Ouest et Europe occidentale,* University of Bordeaux I, doctorat d'État, 1991.

JOHANSON, Donald, and EDEY, Maitland. *Lucy: The Begginings of Humankind,* New York, Simon & Schuster, 1981.

JOHANSON, Donald, and EDGAR, Blake. *From Lucy to Language,* New York, Simon & Schuster, 1996.

JOHANSON, Donald, and SHREEVE, James. *Lucy's Child: the Discovery of a Human Ancestor,* New York, William Morrow, 1989.

LAHAYE, Henri. *Le Lampadaire*, 1995.

MINIATURA, Christian, BAUDON, Jacques, ROBERT, Jacques and GORCEIX, Olivier. *Le Journal du CNRS*, March 1995.

NARA, Takashi. *Étude de la variabilité de certains caractères métriques et morphologiques des Néandertaliens*, University of Bordeaux, doctorat d'université, 1994.

NAZARIAN, Serge. *Contribution à l'étude morphométrique du rachis des hominidés*, University of Aix-Marseille II, doctorat d'État in human biology, 1989.

ORSENNA, Érik. *Histoire du monde en neuf guitares*, Paris, Fayard, 1996.

PELOT, Pierre. *Le rêve de Lucy*, Paris, Seuil, 1990.

PÉNIN, Xavier. *Modélisation tridimensionnelle des variations morphologiques du complexe cranio-facial des* Hominoidea; *applications à la croissance et à l'évolution*, University of Paris VI, doctorat d'université, 1997.

PETTER, Germaine, and SENUT, Brigitte. *Lucy retrouvée*, Paris, Flammarion, 1994.

PICQ, Pascal. *L'Évolution de l'articulation temporo-mandibulaire des hominidés fossiles : anatomie comparée, biomécanique, évolution, biométrie*, University of Paris VI, doctorat de troisième cycle, 1983, and *Cahiers de paléoanthropologie du CNRS*, 1990.

PICQ, Pascal. *Lucy et les premiers Hominidés*, Paris, Nathan, 1993.

PICQ, Pascal, and VERRECHIA, Nicole. *Lucy et son temps*, Paris, Mango, 1996.

PIEL-DESRUISSEAUX, Jean-Luc. *Outils préhistoriques*, Paris, Masson, 1986.

PILARDEAU, Pierre, RICHARD, R., PIGNO, R., MUSSIR, R. and TEILLET, T. *Journal de traumatologie sportive*, 1990.

RAMIREZ-ROZZI, Fernando. *Le développement dentaire des hominidés du Plio-Pléistocène de l'Omo, Éthiopie*, Muséum National d'Histoire Naturelle, doctorat d'université, 1992, and *Cahiers de paléoanthropologie du CNRS*, 1997.

READER, John. *Missing Links*, London, Book Club Associates, 1981.

REBEYROL, Yvonne. *Lucy et les siens*, Paris, La Découverte-Le Monde, 1988.

SCHAEFFER, Pierre. *Faber et Sapiens*, Paris, Belfond, 1986.

SENUT, Brigitte. *Contribution à l'étude de l'humérus et de ses articulations chez les hominidés du Plio-Pléistocène*, University of Paris VI, doctorat de troisième cycle, 1978, and *Cahiers de paléoanthropologie du CNRS*, 1981.

SENUT, Brigitte. *Le coude chez les primates hominoïdes : aspects anatomique, fonctionnel, taxonomique et évolutif*, Muséum National d'Histoire Naturelle and University of Paris VI, doctorat d'État, 1987, and *Cahiers de paléoanthropologie du CNRS*, 1989.

TARDIEU, Christine. *Analyse morpho-fonctionnelle de l'articulation du genou chez les primates; application aux hominidés*, University of Paris VI, doctorat de troisième cycle, 1979, and *Cahiers de paléoanthrolopologie du CNRS*, 1983.

TARDIEU, Christine. *Mise au point d'une nouvelle méthode informatisée d'analyse tridimensionnelle de la marche bipède pour l'étude des déplacements des centres de gravité du corps et de ses différentes parties. Application à l'Homme et aux primates non humains*, University of Paris VII, doctorat d'État, 1987, and *Cahiers de paléoanthropologie du CNRS*, 1992.

TRINKAUS, Erik, and SHIPMAN, Pat. *The Neanderthals*, New York, Alfred A. Knopf, 1993.

USSUNET-ZARROUK, Catherine. *Approche du régime alimentaire de l'Homme fossile par caractérisation des stries d'usure dentaire*, University of Paris VI, doctorat d'université, 1991.

VAN DEN BERG, Gerit Dirk. *The terrestrial faunal successions of Sulawesi, Flores and Java during the Pliocene-Quaternary, and their bearing on the reconstruction of the palaeozoo-geography of elephantoid and hominoid dispersal into the Indonesian region*, University of Utrecht, Ph.D., 1997.

VILLEMEUR, Isabelle. *Étude morphologique et biomécanique du squelette de la main des Néandertaliens. Comparaison avec la main des Hommes actuels*, University of Bordeaux I, doctorat d'université, 1991, and *Cahiers de paléoanthropologie du CNRS*, 1994.

ZEITOUN, Valéry. *Cladistique et paléoanthropologie : le cas de l'espèce* Homo erectus *(Dubois, 1984)*, University of Bordeaux I, doctorat d'université, 1996.